Joseph Glynn

Rudimentary Treatise on the Construction of Cranes and Machinery

Joseph Glynn

Rudimentary Treatise on the Construction of Cranes and Machinery

ISBN/EAN: 9783744678926

Printed in Europe, USA, Canada, Australia, Japan

Cover: Foto ©berggeist007 / pixelio.de

More available books at **www.hansebooks.com**

RUDIMENTARY TREATISE

ON THE

CONSTRUCTION OF CRANES,

AND

MACHINERY

FOR RAISING HEAVY BODIES, FOR THE ERECTION OF
BUILDINGS, AND FOR HOISTING GOODS.

By JOSEPH GLYNN, F.R.S.,

MEMBER OF THE INSTITUTION OF CIVIL ENGINEERS, ETC.

HONORARY MEMBER OF THE PHILOSOPHICAL SOCIETY,
NEWCASTLE-UPON-TYNE, ETC.

FOURTH 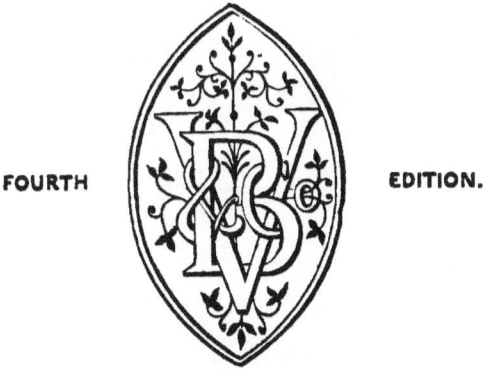 EDITION.

LONDON:
VIRTUE BROTHERS & CO., 26, IVY LANE,
PATERNOSTER ROW.
1866.

CONTENTS.

	Page
Preface	v
1. Primitive State of Man	1
Origin of the windlass	3
Improvements on the windlass	4
Original capstan described	8
Original improvements on the capstan	9
Jack-roll	12
Gin, described	12
Application of steam power to collieries	13
Extensive use of the capstan by the Russians	14
Cross	16
Hand wheel	16
Cog and round	16
Lever and axle	17
Block and pulley, origin and use	18
"Dead-eye"	19
2. Application of Hoisting Machines, and Employment of the Power of Men	20
Derrick	20
Henderson's derrick	21
Triangle, or "three legs,"	22
Calculations, and experiments on the power of men	22
3. Progressive Construction of Cranes	26
Walking crane	27
Goods crane	28
Shipwrights' crane	29
Influence of cast iron on the form of the crane	33
Brakes	33
Well cranes	36
Wharf cranes	36
Construction of the crane post	36
Foundry cranes	41
4. Traversing, or Travelling Cranes	43
Rennie's traversing crane	43
Improved travelling cranes	48
Use of, in the erection of large buildings, &c.	48
Use of, in building bridges	49
5. Self-acting Hoisting Machines	50
Sack-tackle	51
Hoist, or lift	51
Cornish man-machine	52

CONTENTS.

	Page
Hague's vacuum crane	53
Steam crane	53
Armstrong's water crane	58
Murray's lifting apparatus	63
Hydrostatic paradox, explained	64
Bramah's press	65
Bramah's presses, used in the erection of the Conway Bridge	65

6. STRENGTH OF MATERIALS USED IN THE CONSTRUCTION OF CRANES 67
   - Beams, strength and forms of 67
   - Timber beams, experiments on the strength of . . 67
   - Experiments on the strength of cast iron . . . 70
   - Strength of cast-iron pillars 74
   - Tables showing the strength of cast iron under various tests . 77
   - Strength of wrought iron 78
   - Transverse strength of wrought iron . . . 80
   - Crane hooks, form and strength of 81

7. DEFLECTION OF MATERIALS 81
   - Cast iron 81
   - Timber 82
   - Table of the strength of timber in general use . . 85
   - Rules on the strength of timber crane-posts . . 88
   - Lang's masting sheers, their construction described . 90

8. STRESS OR FORCE UPON EACH PART OF A CRANE . . 92
   - Results of alterations, in the form of, considered . . 94

9. CHAINS AND ROPES OF CRANES AND HOISTING MACHINERY . 97
   - Chains, strength of 97
   - Chains for crane work 97
   - Barrels of cranes 97
   - Comparison of the strength of chains and ropes . . 99
   - Huddart's experiments on the strength of cordage . . 100
   - Flat ropes 102
   - Weight and strength of chain and hempen cables . . 103
   - Wire ropes 103
   - Wire ropes, strength of 104
   - Strength of flat wire and hempen ropes . . . 105

10. MACHINERY OF CRANES 106
    - Wheels, strength and proportions of . . . 107
    - Diameters of wheels and barrels 108
    - Toothed wheels 110
    - Strength and form of the teeth 110
    - Willis's method of forming teeth . . . 112
    - Willis's "Odontograph" 114

11. FOUNDATIONS AND MASONRY FOR FIXING AND SECURING CRANES 114
    - Works of reference 116

12. REFERENCE TO THE ILLUSTRATIONS 118

PREFACE.

The subject of this Treatise is the construction and use of machines which diminish toil, and facilitate and lessen labour without superseding it, enabling men to perform what they could not accomplish without such aid.

Deprived of mechanical power, a man's force is limited to his muscular strength, of which he has but little, in proportion to his bulk and weight, when compared with other animals; his disposable mechanical force, when daily exerted for ten hours, being only about one-tenth of his weight.

The old race of millwrights—men who designed and constructed their own work—may be considered extinct; and the operative engineers or "fitters" of modern times, although excellent workmen at the vice or the lathe, have, since the introduction of self-acting tools, and by the classification of labour, become almost machines themselves.

One man has been trained to do one thing, in doing which, however skilful he may be, he exercises no discretion of his own; he has nothing to contrive or to proportion. But good springs out of evil. Providence did not intend that man should be reduced to a machine; his mind will not rest satisfied in this condition; he begins to inquire why he finds himself thus; he desires to know the relation between cause and effect, and to understand the principles on which are founded the orders he is called upon to execute.

It is hoped that the elementary treatises now put forth may serve as guides to such persons, as well as to young students, in commencing the pursuit of knowledge, and tend to render the course straighter, and the task less difficult.

The author has for many years had the direction and management of men in considerable numbers. He is convinced that perfect order, strict discipline, and prompt obedience, are imperatively necessary to ensure success in the combined efforts of many men; but he is also convinced that intelligent and well-informed people are more easily directed than those who are uneducated and ignorant; and he has never found that a sound education, and a right understanding of first principles, unfitted a man for the station he might hold, although they might tend to raise him above it, and often eventually did so. He has the gratification of seeing many persons, who have acted under his orders, now filling offices of trust and responsibility with merited credit, and others deservedly acquiring reputation and wealth, which they owe to the early cultivation of their minds.

In every state of society the many must be ruled by the few, and "those who think most must govern those who toil;" but the relations of society in this country at present have the effect of increasing wealth in few hands, and many men labour to make one man rich.

This may, in part, be attributed to the use of machines, as substitutes not only for labour, but for the performance of operations formerly requiring skilful workmen.

Machines are employed to make machines, and thus capital increases in a larger ratio in few hands.

It may be doubted whether the accumulation of capital in large masses be a national benefit. If it be otherwise, if it be desirable to improve the condition of "those who have most of the toil and least of its benefits," as has been well said by an illustrious Prince, then will one of the best and most peaceful means of modifying this unequal distribution of comforts be to give to the working-classes a sound and useful education, and to impart to them a knowledge of first principles in the mechanical arts they are called upon to practise; to elevate their character, and better to fit them not only to fulfil their duty in that station wherein Providence has placed them, but to render them capable of rising above it, when opportunities are presented to them, by peaceful and legitimate means, conducive to the general welfare of society.

J. G.

ON THE

CONSTRUCTION OF CRANES.

Primitive Methods of Assisting or Combining the Power of Men or Horses.

The Windlass.—The Capstan and the Gin.—The Cross.—The Handwheel. —The Cog and Round.—The Dead-eye.—The Pulley.

If man were furnished with no other means of defence, or of assistance to his physical strength, than those which his own organisation supply, he would be one of the most helpless creatures existing. But his hand instinctively grasps the club or the stone as ready weapons for his protection; and his further wants, stimulating his ingenuity, teach him to form the objects within his reach into the bow, the spear, and other appliances for the pursuit of game or of fish. He twists the vegetable fibre or the thong into the line or the cord, and the cord into the rope. From the fallen tree he makes the raft and the canoe. He quits the cave which gave him shelter, and builds the more convenient hut and the more ample cottage, and he soon finds that he has to deal with materials beyond his unaided strength to fashion or to move.

The pole in his hands becomes a lever to remove the trunk of the fallen tree, and the rope of twisted thongs or fibres of bark, thrown over the fork of an extended branch, probably formed the first crane. Although no mechanical power be gained by this arrangement, yet it enables several men to unite their strength, and one man to maintain and hold fast the result of their combined efforts.

That either this form, or the rude but useful adaptation

of the lever, common in all parts of Northern Germany,
Fig. 1.

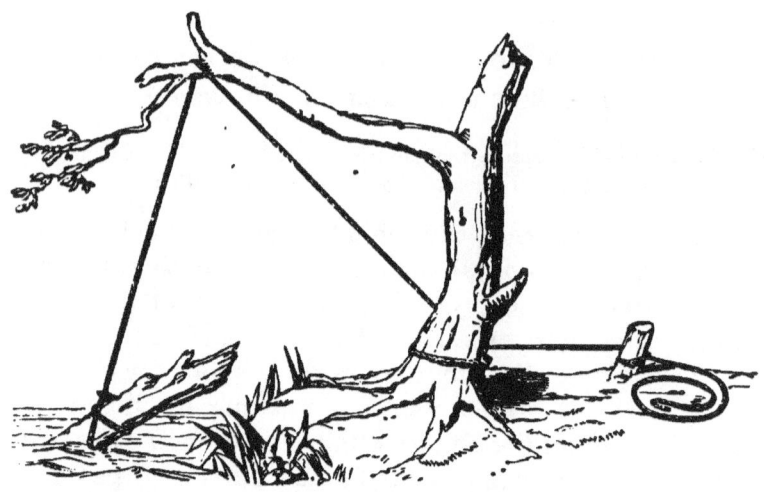

and sometimes seen in a more slender and commonplace shape in our own brickyards, may have been the original crane, seems not improbable; the name given to the machine being the same as that of the long-necked water-fowl, which, wading in the shallows by a river's side plunges its bill into deeper pools to bring up its food. The Crane, la Grue, der Kranich, la Grua, and Cigoñal, indicating in French, German, Italian, and Spanish, that the idea which furnished the name was the same in them as in the English language, or that they all derived the machine and the name from the same source, probably from the Germans.

Most persons who have passed through German villages will remember the simple and picturesque mode of raising water from a well by means of a tall fir or poplar tree resting in the fork of an elm growing near the well or brook.

The root end of the poplar, assisted perhaps by the weight of a stone, overbalances the top, from which the bucket is suspended; the counterbalance being equal to half the weight to be raised, or thereabout, so that the man has to pull down the bucket to make it descend into the well; the counterbalance assisting him in hoisting up the bucket when full; and thus, by apportioning his efforts, he doubles his effective force at the time he needs it.

Fig. 2.

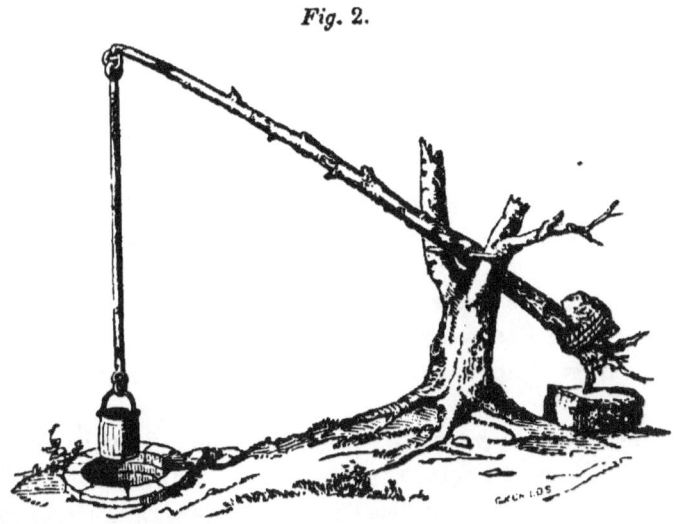

The application of the pole as a lever for moving weights, or for turning over the trunk of a tree, might suggest its further use, combined with a rope, to obtain mechanical power; the author has seen old seamen, in case of need, resort to such a contrivance, and make, as they termed it, "a purchase," with a plain wooden roller (part indeed of a fir-tree trunk), a rope, and two handspikes, and, by the skilful combination of these materials over a hatchway, accomplish what the united strength of all hands failed to effect without such assistance.

A little in advance of this, is the mode of hoisting timber common throughout all the North of Europe, and sometimes used by our woodmen in England, by means of the "lever and axle" attached to the "three legs or triangle"—a tripod, formed of three poles, secured by a rope or shackle at the top. The end of one handspike or spoke, being occasionally thrust through the axle or windlass, rests upon the ground and stops it from unwinding, forming a simple but effectual check.

This same primitive windlass is still used by the Chinese for weighing anchor, even in their largest junks. A strong tree of hard wood extends entirely across the vessel from side to side; it is hewn into an octagonal form, and the ends being reduced and rounded to form pivots, rest in the top timbers of the ship, which are brought up above the deck for this purpose; consequently the length of their windlass is sometimes 28 or 30 feet. It is stopped from unwinding or turning the wrong way by thrusting a handspike through, and allowing its end to bear upon the deck, whereby the descent of the anchor, between the successive exertions of the crew, is prevented, until they can ship their bars and heave again.

The next improvement appears to have been "the paul."

So late as the year 1776, when Falconer, author of that beautiful nautical poem, the Shipwreck, published his Marine Dictionary, the windlass used on board of British merchant vessels appears to have advanced but little beyond this primitive form. The differences were two: first, it did not rest on the ship's sides, but was supported and secured in two strong timbers fixed on opposite sides of the main deck, a little behind the foremast, wherein the windlass turns on its axis. These are generally called "the windlass bits," and are each made in two pieces, for more conveniently getting out the windlass and allowing the bight of the cable to be passed round it, as it commonly is in three turns; the upper parts of these bits being formerly ornamented with carved "knights' heads," still retain that name. Secondly, the windlass was furnished with pauls, which Falconer thus describes:—

"The pauls, which are formed of wood or iron, fall into notches cut in the surface of the windlass, and lined with plates of iron. Each of the pauls being accordingly hung over a particular part of the windlass, falls eight times into the notches at every revolution of the machine, because there are eight notches placed on its circumference under the pauls. So, if the windlass is twenty inches in diameter, and purchases five feet of the cable at every revolution, it will be prevented from turning back, or losing any part thereof, at every seven inches, nearly, which is heaved in upon its surface.

"As this machine is heaved about in a vertical direction, it is evident that the efforts of an equal number of men acting upon it will be much more powerful than on the capstan, because their whole weight and strength are applied more readily to the end of the levers employed to turn it about; whereas, in the horizontal movement of the capstan, the exertion of their force is considerably diminished. It requires, however, some dexterity and address to manage the handspike to the greatest advantage; and to perform this, the sailors must all rise at once upon the windlass, and fixing their bars therein, give a sudden jerk at the same instant, in which movement they are regulated by a sort of song or howl pronounced by one of their number."

The "song" of the seamen, when raising the anchor for their departure, has always a melancholy and plaintive tone, even

"When ten jolly tars, with musical Joe,
Heave the anchor a-peak, singing, Yo, heave ho!"

The windlass remained nearly in the same state as described by Falconer, when the author first saw it; but the windlass necks were then made of iron, so as to prevent loss of power by friction, and the pauls, then placed in the centre of the windlass, were two in number, and of different lengths, or, as the sailors termed them, "paul and half-paul," which, although the number of notches was still only eight, had the effect of dividing the circumference of the windlass into sixteen, and enabled the seamen to retain every three and a half inches of the cable they hove in. This was a great relief to the men, for seven inches on the windlass occasionally lost by the shock of a heavy sea, occasioned a severe jerk at the end of a six-feet handspike.

The advantages arising from this improvement gave rise to the patent pauls, wherein a cast-iron paul-wheel was fixed upon the wooden windlass, and the pauls were made more numerous, and of various lengths. The paul-wheel had sixteen notches; it was made hollow, and octagonal inside, to slip over and be adapted to the old windlass; better methods of securing all the parts being introduced at the same time, the announcement in the Newcastle papers became less frequent, that " the Good Intent, of Shields (coal-laden), with windlass upset, and loss of anchors and cables in the Swin, had put into Harwich, and must discharge cargo for repair."

The principle of dividing a ratchet-wheel by differential pauls, whereby minute division of circumference can be obtained with a comparatively coarse pitch of tooth, is worthy of being borne in mind, and may be made useful in many kinds of machinery. By applying it to the ship's windlass, every inch of cable is retained, much labour is saved, and the men are preserved from hurts, often caused by the violent recoil of their handspikes.

Other alterations have been successively made in the windlass; machinery of various kinds has been attached to it, for some of which patents have been granted.

One of these alterations consisted in fixing within the bit-heads a sway-beam of wrought iron, constructed so as to be unshipped at pleasure. Upon the windlass, and immediately under the two arms of the beam, were fixed ratchet wheels, upon which two pauls, one attached to each arm, alternately acted as the beam was raised or depressed, and a wooden pole or handle being passed horizontally through an eye at each end of the sway-beams, they were worked with a pumping action like a fire-engine. This arrangement rendered the working of the windlass continuous, or nearly so, as well as much more rapid in its action, and highly useful for lighter work, such as warping the ship out of a crowded harbour; whilst, in case of need, it might be readily removed, and the windlass worked with handspikes as before. By a subsequent improvement, patented by Messrs. Pow and Fawcus, of North Shields, the ratchet

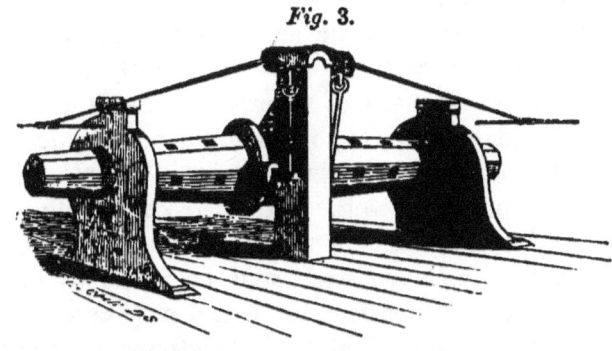

Fig. 3.

a, a, are wheels with plain surfaces; there are two of these "purchase-wheels" fixed upon the windlass.

c, c, the nipping levers confined to the purchase-wheels by flat iron rings, or discs, called travellers, which, in descending, move freely round.

d, d, the travellers with the cheek-plates of the nipping levers bolted to them; the levers bite upon the purchase-wheels, and, acting alternately, force the windlass round in their ascent.

e, the cross-head of the sway-beam shown in plan, with sockets to receive the levers.

f, f, the levers, which are bent forward to clear the knees of the bit-heads.

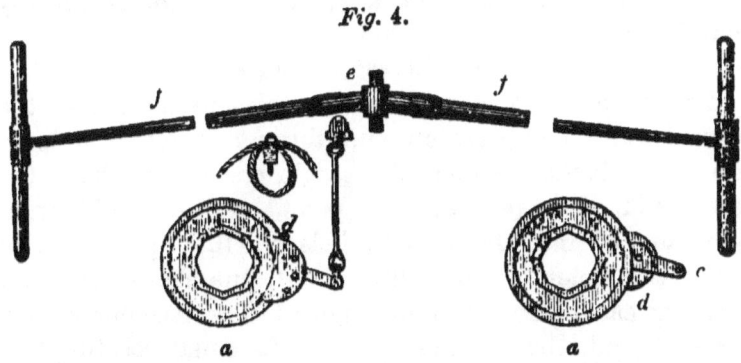

Fig. 4.

wheels are dispensed with, and a nipping lever, acting on a wheel with a plain surface, is substituted, so that the noise of the ratchets is avoided, and all the length of cable hove in is held fast. Other alterations consisted in the application of toothed wheel-work in various ways, sometimes to the ends, and sometimes to the middle of the windlass, so that at last it has, in fact, become a powerful compound piece of crane-work.

The ancient mariner, however, looks on all these alterations with jealousy and suspicion, prudently rejecting all such as may not, in case of emergency, be laid aside for the trusty handspike of tough ash or hiccory, which never has its teeth broken by a heavy strain in a gale of wind, like the cast-iron wheel-work. He also knows the importance of having such tackle as the ship's carpenter may repair, and for which materials may be found in any port whither, in stress of weather, his vessel may be driven.

As it is often inconvenient, and sometimes dangerous, to ship and unship the handspikes in a ship's windlass, besides causing much loss of time, and as by it the united strength of many men cannot be employed, the capstern or capstan is used instead of it, in large vessels, to weigh the anchor, and in ships of war, when despatch is needful, a large body of men act together, walking round the capstan, their efforts being rendered simultaneous and uniform by the sound of music, and the cable of the gallant ship, on her return home from a foreign station, is merrily rounded in.

"A fair wind, and off she goes."

The original capstan or crab was, something like the primitive windlass, set on end through a round hole in the deck; it was formed of a single piece of timber; the lower end of it, reduced to a pivot, and shod with iron, was stepped into the vessel's kelson; the upper part of it had two holes morticed through it, one above the other, crossing each other at right angles. The machine in this form may still be occasionally seen in small coasting vessels from the bye ports in the West of England, or on board of French luggers; and it is sometimes used for hauling up the larger fishing craft upon the beach. It gradually, however, assumed a more important shape, and instead of being only one piece of wood, it was composed of several parts; namely, the "drum-head," the "barrel," the "whelps," and the "spindle," all, in the first instance, made of timber. The spindle, as before, was shod with iron, to form the pivot, and worked through a round hole; but this was formed in a strong wooden stock, called "the step," which rested upon and was bolted to beams placed for the purpose; the spindle was hooped with iron to prevent abrasion of the wood, and it revolved in an iron socket or collar, called "the saucer," which was fixed in the step. Two strong pauls of wood or iron were bolted at the deck to the beams above mentioned, and acting against the lower end of the whelps, prevented the recoil of the capstan.

A man-of-war has two capstans; the main capstan being, as it were, a double one, like two capstans on the same spindle, the one on the main deck and the other on the upper deck, so that two tiers of bars can be worked at once upon the two decks. Many improvements have also taken place in the capstans; the spindle is now made entirely of iron, and, wheel-work being applied to it, the capstan also has become a compound machine, displaying in some instances much ingenuity.

In this sketch, the drum-head is fixed upon the spindle, and turns it round. An iron bolt passing through the drum-head, locks it to the barrel, and the whole capstan turns round with the spindle, forming a simple machine, or "single purchase." When the locking bolt is withdrawn, the wheel-work, *shown in plan*, acts between the spindle and the barrel, and a power of three to one is gained; the spindle makes three turns, while the barrel makes one, and they revolve in opposite directions. The mode of locking the drum-head is shown in the sectional elevation of the capstan.

The mechanism, has, however, this general character in all its phases, namely, that there is a toothed pinion upon

Fig. 5.

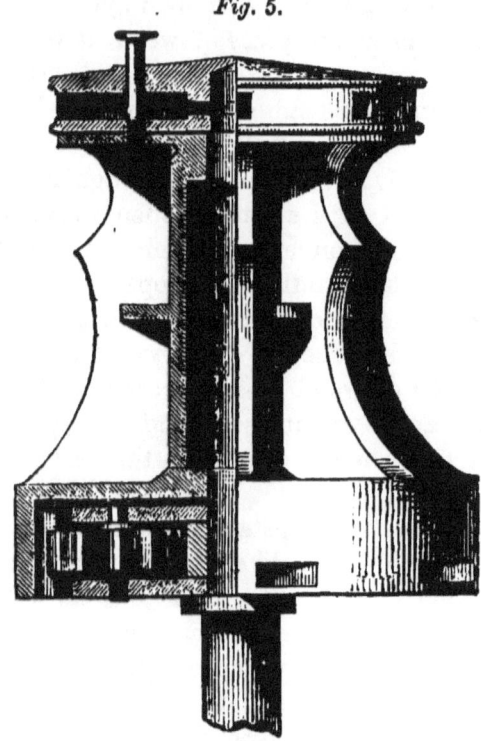

the spindle, and a wheel with external teeth attached to the lower part of the barrel.

Fig. 6.

ON THE CONSTRUCTION OF CRANES. 11

The working of capstans is subject to an inconvenience, arising from the tendency of the rope which is wound upon them to advance in a spiral direction towards the end of the barrel, there being generally two and a half or three turns of the rope taken round the capstan, and one part of the rope is wound off the barrel as the other part is wound on. It therefore becomes necessary to "surge the capstan," that is to say, suddenly to slack the rope so as to bring it back to its original position; the barrel and whelps being formed like a truncated cone to facilitate the operation.

But this being always inconvenient, and often hazardous, especially where very great weights must be dealt with, a plan has been devised which may be seen in use at the great masting shears in Her Majesty's Dockyard, at Woolwich, to obviate this objection. Two capstans are fixed near to each other, and are connected by a pair of toothed wheels, so that they revolve together, but in contrary directions; one only has a drum-head to receive the capstan bars, the other is made low to allow the bars to pass over it, and the rope is passed round both in the form of the figure of 8.

The crab or capstan, on a large scale, is often used at mines and coal-pits to raise and lower the heavy cast-iron pumps, to draw the pump-rods, and other similar work; it has generally four arms permanently fixed, instead of bars, to which a large number of the miners apply their strength, and they are sometimes assisted by horses.

The mine capstan is a simple, and, in early times, has been a very useful machine; but, since the mines have been sunk deeper, it often causes frightful accidents: the men not being accustomed to act in concert like the well-trained crew of a man-of-war, and being hastily called together to raise great weights of pumps, or, what is more dangerous, to lower them, are occasionally overcome by the descending load, and this accelerating in its descent, "the capstan spins," flinging the miners from its arms with fearful violence, when broken limbs and even death ensue. In these days, when safe and powerful machinery may be substituted, the cheapness of the mine capstan is the chief inducement to retain it.

Wherever the author could exercise sufficient influence, he has erected a combination of wheel-work for deep pits instead of the mine capstan; so that a small number of men suffice, and accidents are avoided. There is scarcely any limit to the powers that may be gained by compound wheel-work, and the additional time it takes is of little moment in heavy mining operations.

During the early periods of mining in Cornwall, when pits began to be sunk in those places where the manifest abundance and richness of a vein led the miner to penetrate beneath the surface, the produce was raised by being thrown upon successive stages or platforms, or "shammels," as the Cornish miners call them, by men stationed at different elevations. The introduction of the jack-roll or windlass, in its rudest shape, was a great improvement; this contrivance, which probably came from Germany, not only facilitated the raising of the excavated material, but enabled them to clear the mine of water by means of buckets, with a degree of despatch not before practicable.

In the Derbyshire lead mines, the jack-roll or "wallower," as they term it there, is still used, as it is also at some of the ironstone pits; but at many of these, as in the collieries, it has been superseded by "the gin," which is worked by horses.

The gin consists of an upright wooden axle, on which is fixed a hollow cylinder of wood-work, called the cage, round which a rope winds horizontally; the ends of the rope being directed down the pit by two pulleys. A transverse beam, seven or eight yards long, is secured across the axle, to each end of which is yoked a horse. The horse track should not be less than seven or eight yards in diameter, so that the horse may not expend his force in an oblique direction, but get a fair pull on the "starts."

An ordinary horse produces the greatest mechanical effect in working at a gin, or drawing a load on a tramway, when he travels at the rate of two and a half miles an hour or 220 feet in a minute; he can then exert in regular work day by day for eight hours, a steady pull of 150 pounds. Hence arises our familiar term of "horse's power;" the speed of 220 feet multiplied by 150 pounds being equal to 33,000 pounds raised a foot high in a minute, a standard which engineers have by common consent adopted as the expression for mechanical powers employed for practical and manufacturing purposes.

This limit to the use of horses for winding coals, the adaptation of the steam engine to a rotatory motion, and its successive improvements having made it perfectly manageable, the increasing depth of the pits, and the demand for coal, have caused the steam-engine in most cases to supersede the gin, and its use has enabled the coal-owners to expend large capital in sinking deep pits, where thick and valuable coal might be obtained; so that the power of the engines has, of late years, been continually augmented.

A company of gentlemen, having recently sunk a pit to a depth of 220 yards, at Cinder-hill Colliery, near Nottingham, with some difficulty and much expense, found coal of good quality and ready sale; and, being desirous to meet the demand in the market without sinking additional pits, they applied to the Butterley Company to furnish them with such an engine as might raise what coal they could sell. The author was desired to make the requisite calculations, and to prepare such plans as should effect this object. He found that it would be necessary to draw from this depth a ton of coal in every successive minute for ten hours a day; that the coal should be brought up in two of the small underground waggons, to be unloaded "at bank," each of which contained half a ton; and that, in order to allow time to land each load and to reverse the engine, it was requisite to bring up the ascending "cage" as fast as the descending one would with safety and certainty go down. Taking into account the resistance of the air and the friction of the "guides," it was not considered prudent to attempt a speed greater than 16 feet in a second. A heavy body, unchecked by the friction of slides or guide rods, falls about 16 feet in the first second; but as the engine, when reversed, does not immediately regain full speed, it permits the requisite acceleration of the falling or empty cage, which descends as the full one rises. The weight of the flat rope, the friction of the cage, and the weight to be drawn at such a speed, rendered it necessary to construct a non-condensing steam-engine of 200 horses' power. The result was quite successful; and, in the collier's week of five days, or 50 hours' work, 3000 and sometimes 3500 tons of coal are raised with ease. Much more than this is performed at several of the collieries in the counties of Northumberland and Durham, where engines of great power are employed at larger pits.

Referring to the practice of heaving at the capstan on board of ship to the sound of music, it may be remarked that, by this means, a number of those machines, actuated by large bodies of men, may be made to exert their force at once upon the same object; and that the Russians of the present day employ them in moving those immense blocks of stone, of which their public buildings display so many examples; and, also, that they are employed in moving their line-of-battle ships, often built on shallow water at a distance from the sea, until they are fairly floated upon the caissons or "camels," which are used to buoy them up and enable them to come down the Neva to the Gulf of Finland, towed by a flotilla of row-boats.

The rock on which stands the colossal statue of Peter the Great, was moved from Lachta, in Finland, to the Russian capital by the aid of many capstans worked at the same time by a large body of soldiers, who kept step to the sound of the drum. The impression which the sight of this immense monument made on the author's memory, many years ago, is still fresh and vivid.

The rock, when brought to St. Petersburgh, is said to have weighed 1100 tons, which corresponds with the original dimensions of the stone. These were 42 ft. long at the base, 36 ft. at the top; 21 ft. thick, and 17 ft. high.

The transit of this enormous block of granite was facilitated by a kind of anti-friction railway, laid down as it proceeded onward, and taken up from behind; it consisted of large beams of timber, wherein grooves were formed to receive large cannon-balls, the stone resting upon corresponding grooved timbers, so that the two beams formed a kind of channel for the balls.

The taste of the sculptor unfortunately led him to dress the stone, and partially to change its form, by which its size was reduced and its rude grandeur impaired, to the great chagrin of the Empress Catherine; but still it is a noble work.

Few monuments can be compared with the Bronze Statue at St. Petersburgh; the animated figure of the rampant horse, standing 17 ft. high, with his imperial rider 11 ft. in height, admirably designed and skilfully executed, poised upon their massy pedestal, produce an effect hardly to be surpassed.

It is much to be regretted that some of the best machine capstans that have been introduced on board ship have failed from the weakness of their wheel-work and the imperfection of their workmanship, thus creating a prejudice among seamen to such mechanism, and tending to delay, if not to prevent, its more extended use. In all such cases the stress should be ascertained, and the strength of the machinery calculated, making allowance for contingencies, instead of taking it for granted, as is too often done, that wheels and teeth of a certain size and pitch will answer the purpose. The constant inconvenience and frequent danger incurred in shipping and unshipping the handspikes, or spokes, caused them, in small windlasses, to be permanently fixed, and eventually the cross or the hand-wheel became, for many purposes, parts of machines, as we see them still on the copperplate printing rollers and the steering wheels of ships, by which mechanical power is gained in the proportion of the radius of the spoke to the semi-diameter of the barrel.

The same rule holds good in all the simple forms of the wheel and axle, whether windlass, capstan, jack-roll, or gin.

The spokes of the hand-wheel were generally eight in

Fig. 7.

number, but it was not difficult to perceive, that if their number were increased, or if a second wheel and axle could be brought to act upon the first, much greater power might be gained. Hence "*the spokes*" were multiplied into "*cogs*," and upon the second axle was fixed a small wooden wheel or "*lantern*," composed of two discs or trenchers, in which were inserted six or more staves or "*rounds*" of hard wood like the rounds of a ladder. The figure here given was taken from an old deep well in the county of Kent.

Machines of this kind were common forty years ago, and some may still be in existence, although the general introduction of cast-iron machinery and its toothed wheels has now superseded the old "cog and round."

The mechanical power gained by the wheel and axle, or level and axle, which are the same things, has been already mentioned; there is, however, another form of the axle, by which much greater power is gained for a short lift. It is obvious that the only way to gain more power with the simple lever and axle, is to increase the length of the lever; but this can only be done within a very limited range.

Fig. 8.

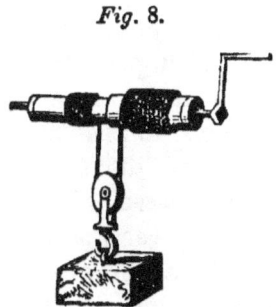

If, however, the axle be made of two different diameters, one-half of the barrel's length being a little larger in diameter than the other half; if a single pulley or block be put upon the middle or "bight" of a rope, and the two ends of the rope be wound round the two ends of the barrel in opposite directions, so that one end will wind off as the other end winds on; then, if the smaller part of the barrel be 7 in. in diameter, or 22 in. in circumference, and the larger part be 8 in. in diameter, or about 25 in. in circumference, every turn of the barrel will wind up the difference, or about 3 in. of rope; which difference, being divided by the use of the pulley, would raise the weight suspended to it about an inch and a half, so that with a winch or lever of 18 in. radius the workman's hand would move in a circle of 3 ft. in diameter, or about 113 in.

in circumference; and consequently a power of about 75 to 1 would be gained. But, as one complete turn of the barrel winds on 25 inches of rope, it requires nearly 17 ft. of rope to raise the load 1 ft. high. The quantity of rope required limits the use of the differential axle or barrel to a short range; but there are many cases in which so simple and powerful a machine may be very usefully employed.

In the machine last described the block or pulley formed an essential part, and added to the power of the machine. As it is used in the construction of many cranes as a means of gaining mechanical power, it may be well to show, briefly, how it does so.

When ropes or cordage came into use, it was found to be a convenient mode of raising any round object up a slope or inclined plane from the water-side, or of lowering it down in similar situations, to pass the middle of the rope about a tree or a post, and the two ends of it round the object to be raised or lowered. In this way a heavy spar may be hoisted from the river upon a quay or wharf, or a water cask lowered into a ship's boat with comparative ease, the power gained being two to one, independent of the inclined plane, the cask itself serving as a pulley. This arrangement of the rope, by sailors termed a "parbuckle," is also used by the draymen in London with great skill and dexterity; they will sometimes lower a cask of half a ton in weight into a cellar without any apparent difficulty, by making it form a part of the mechanism.

The combinations of ropes and rollers to gain mechanical power, which invention sharpened by necessity would soon suggest, led to the contrivance of "*the block*," which at first was merely a piece of hard wood with a hole in it, to reeve the rope through, such as are still used to "set up" or heave tight the shrouds and standing rigging of ships. These blocks have three holes in them, through which the rope or "laniard" is passed, and then greased, to reduce the friction, until it is hove tight and made fast. The round shape of the block, and the position of the three holes, give it somewhat the resemblance of a death's head, and hence its name, "the dead-eye." In blocks constantly used, the friction and rigidity of the cordage causing so great a loss of power, induced the addition of a roller or "sheave" in the block itself; and successive improvements have brought the crane blocks, used in some of the steam-engine manufactories, to their present state of excellence. The blocks now employed in the leading establishments of the present day, with iron shells, brass sheaves, and steel pins, their wrought-iron straps or side links, swivels, and hooks all carefully calculated, so as to give sufficient strength of materials for the load to be hoisted, and at the same time to avoid superfluous weight, and to reduce the friction by proportion of parts and superior workmanship, greatly diminish labour, and increase despatch in the manufacture of heavy machinery. The introduction of iron blocks and pulleys admits the use of chains instead of ropes; and as the links of chains are now made almost exactly uniform in size and shape, they have been substituted for ropes in most foundries and engine works.

The power gained by any combination or system of blocks or pulleys, is proportionate to the distance travelled by the moving force, compared with the height to which the load is raised in the same time, without deducting loss by friction; so that, if mechanical power of two, four, or six to one be gained, the force applied must move through so many times the space that the weight is lifted, and in the same time; for in this, as in all other machinery, speed is lost in proportion to power gained, besides the loss arising from friction of the mechanism and other resistances; consequently, no force descending can ever raise an equal weight to the same height in the same time.

Simple and obvious as these things seem, they have been too often forgotten, and much time and money have been spent in contriving and constructing complicated machines to no purpose.

Application of Hoisting Machinery, and Employment of the Power of Men.

Having traced the early development of mechanism for raising or lowering heavy bodies, its application to practical purposes in different situations must next be considered.

In this there are several points to be determined, namely, the weight to be raised, the height to which it must be hoisted, and the time in which it must be done must also frequently form an element in the calculation, to determine the power to be employed, the machinery to be used, the mode of fixing or attaching the machinery, and of suspending the weight.

When the weights to be raised are those of ordinary merchandise, to be hoisted from the hold of a trading ship, and lowered into a barge alongside, it is usual to raise a single pole, frequently a spare topmast or boom, and to step it over end, immediately before the mainmast, and inclining over the main hatchway of the vessel, or, in sailor's phrase, "to rig a derrick."

The foot of the derrick is stepped into a piece of wood secured to the deck and hollowed to receive it, and the heel of the derrick is provided with "a lashing" of rope to prevent the foot from tripping. The head of the derrick is furnished with a strong rope called "the stay," the end of which is made fast to the head of the mainmast; and there are also two other ropes called "guys," made fast to the head of the derrick, and thence extending one to each side of the ship; so that, by hauling in the one guy and slackening the other, the derrick is made to turn so far upon its heel, and the head, with the load suspended from it by a pulley or blocks, describes the segment of a circle from the hatchway of the ship to the barge which receives the goods. The winch or barrel which winds up the rope is commonly attached to the fore part of the mainmast, and as the current weight of merchants' goods is seldom more than a ton, as, for instance, a sugar hogshead, this arrangement is found very convenient and useful.

A mode of combining the advantages of the derrick with those of the crane has been patented by Mr. Henderson. The jib of his crane is fitted with a joint at the foot, and has a chain instead of a tension bar attached to it at the top, so that the inclination of the jib, and consequently the sweep or radius of the crane, may be altered at pleasure. A similar crane, in a rude form, has long been used in stone quarries; but Mr. Henderson has introduced a parabolic barrel, similar to the fuzee of a watch, upon which the chain winds as it raises the jib, and the barrel decreases in diameter as the jib approaches the horizontal line, so that the power to raise or depress the point of the jib is equalised at all times.

When the weights to be lifted are heavy, and the height to which they must be raised is considerable, as in the masting of a ship of war, or in placing the boilers on board of a steam-frigate, it is customary to employ two strong spars set apart at the foot, but meeting together at the top in an acute angle, where they are secured to each other by a rope lashing, or for permanent purposes by an iron bolt and shackle, from which the requisite blocks are suspended. These spars are stepped, like the derrick, near to the edge of a quay or wharf, or upon the gunwale or side of some large old ship or hulk; from their crossing each other when lashed together at top, something like a pair of large scissors, or "shears," they have received and still retain that name, although in most modern examples no resemblance remains to the original shears. The weights to be lifted at the royal dockyards have of late years become so heavy, and the bulk of the materials, such as the boilers of the war steamers, so great, as to render the employment of the "shear hulk" inconvenient; and permanent shears have, therefore, been fixed upon the quays, and those at Woolwich Dockyard are a good example of the kind. The shears, however, have only one motion in their step, which serves as a centre; the spars, as the radius, describe a vertical arc, inclining their heads over the water, whereas the derrick has two motions, and can describe a vertical and a horizontal arc also. The machinery used with the shears is generally a powerful capstan.

When the weight requires to be lifted perpendicularly, and the height is not great, as, for instance, when some massy stone or beam of timber, iron girder, or the like, must be raised from the ground, so that a waggon may be run under it for convenient loading, it is usual to employ three spars, meeting at the top and spreading asunder at the foot, an arrangement which workmen call the "three legs" or "triangle." In this instance no motion can be obtained beside that of the perpendicular lift. It is, however, a very useful arrangement, easily made in most situations, often enabling a few men and horses to load and remove the largest timber trees and blocks of stone which their waggon is capable of carrying. The machinery used in such cases when men's power is applied, is generally a windlass and a pair of threefold blocks, the windlass being fixed to two of the "legs." When horses are employed, the rope from the blocks or "tackle fall" is passed through a leading block or "snatch block" attached to one of the legs, in order to give the rope a horizontal direction, and the horses being yoked to it, gaining by the threefold blocks a power of six to one, can raise great weights with much facility. The horses regularly engaged on such work display great sagacity and obedience to a word or sign, to hoist, to lower, or to stop.

The power of horses has been already mentioned; the power of men is next to be considered. The late Mr. John Walker, an able assistant of Mr. Rennie, made many and repeated experiments on the power of men employed in raising weights for driving piles in the Royal Dockyard at Sheerness, and he found that the force exerted by an ordinary labourer, in average daily work, frequently did not exceed 12 lbs., and that 14 lbs. was as much as could be reckoned upon as the power of a labourer working daily at a winch or crane handle, for ten hours a day, moving at the rate of 220 ft. per minute.

It is important to remember facts like this, because most writers rate the power of men much higher. This is an error into which they were likely to fall when manual force was exerted for the purposes of experiment, for a short period, or even for a single day. Mr. Joshua Field (late President of the Institution of Civil Engineers) some years ago tried a series of experiments on the strength of men working at a crane of the usual construction, in ordinary use, and not prepared in any manner for the experiments, having two toothed wheels of 92 and 41 cogs, and two pinions of 11 and 10 cogs; the diameter of the barrel, measuring to the centre of the chain, was $11\frac{3}{4}$ in., and the diameter of the circle described by the crane-handle was 36 in.; the ratio of the weight to the power by this combination was 105 to 1.

The weight was raised in all cases through $16\frac{1}{2}$ ft., and so proportioned in the different experiments as to give a resistance against the hands of the men of 10, 15, 20, 25, 30, and 35 lbs., *plus* the friction of the apparatus.

The resistance occasioned by the friction of the apparatus being a constant element in all machines, and of much the same amount in most cranes, and the object being to obtain some practical results on the power of men in raising weights by a system of machinery, it was not thought necessary to make any experiment for ascertaining the amount of this resistance in the present instance.

The following table shows the resistance at the handle, the weight raised in each experiment, the time in which the weight was raised, and the remarks which were made at the time with respect to the men. A column also expressing the power or effect, by the number of pounds raised one foot high in one minute, is added. It will be necessary to add a few words respecting the construction of this column.

In order to compare these experiments with each other,

the results must be reduced to a common standard of comparison, and it is very convenient to express the results of such experiments by the pounds raised one foot high in one minute, this being the method of estimating horses' power. The number is in each case obtained in the followin manner. Take the first experiment.

Here 1050 lbs. were raised 16½ ft. high in 90 seconds; this is equivalent to (1050 + 16·5 =) 17325 lbs. raised 1 ft. high in 90 seconds, which is equivalent to (17325 ÷ 1·5 =) 11550 lbs. raised 1 ft. high in one minute.

In this case the man's power is equal to 11550, and the same calculations being pursued in the other cases, give the numbers constituting the last column in the following

TABLE.

No. of the Experiment.	Statical Resistance at the handle.	Weight raised.	Time in Seconds.	Time in minutes.	REMARKS.	Man's Power.
I.	10	1050	90	1·5	Easily by a stout Englishman	11550
II.	15	1575	135	2·25	Tolerably easily by the same man	11505
III.	20	2100	120	2	Not easily by a sturdy Irishman	17325
IV.	25	2625	150	2·5	With difficulty by a stout Englishman	17329
V.	30	3150	150	2·5	With difficulty by a London man	20790
VI.	35	3675	132	2·2	With the utmost difficulty by a tall Irishman............	27562
VII.	...	...	150	2·5	With the utmost difficulty by a London man. Same as Experiment V.	24255
VIII.	...	...	170	2·83	With extreme labour by a tall Irishman	21427
IX.	...	...	180	3	With very great exertion by a sturdy Irishman. Same as Experiment III.	20212
X.	...	...	243	4·05	With the utmost exertion by a Welshman	15134
XI.	...	...	35	0	Given up at this time by an Irishman.	

Experiment IV. may be considered as giving a near approximation to the *maximum power* of a man exerted for two minutes and a half; for, in all the succeeding experiments, the man was so exhausted as to be unable to let down the weight. The greatest effect produced was that in Experiment VI. This, when the friction of the machine is taken into the account, Mr. Field considered to be fully equal to a horse's power, or 33,000 pounds raised 1 ft. high in one minute. Thus it appears that a very powerful man, exerting himself to the utmost for *two minutes*, comes up to the *constant* power of a horse; that is, the power which a horse can exert for eight hours per day.

Mr. Field's experiments show what a man can do for a short time; Mr. Walker's showed what he can do, day by day, the whole day through. The men employed by Mr. Field were strong and athletic; Mr. Walker's were ordinary labourers, and their power, expressed by multiplying 220 ft. per minute by 12 or 14 lbs., is 2640 or 3080, which must be regarded as the limit of an ordinary man's force constantly exercised at a crane-handle. The author has erected many cranes of various kinds for various purposes; and he has found, practically, that although a man may exert a force of 25 pounds for short periods, yet it is not prudent to reckon upon more than 15 pounds in constant action upon a crane-handle moving at the rate of 220 ft. in a minute. The power of a man will, in that case, be represented by $(15 + 220 =)$ 3300.

Mr. Smeaton, in one of his reports, directs that the water left in the dock at Port Glasgow, which at a medium tide amounted to 2141 cubic feet, or $627\frac{1}{4}$ tons, shall be pumped out by manual labour; and, in describing the pumps, he says:—" This quantity to be raised in four hours to the mean height of 4 ft., will require six men working at a time; and good *English* labourers will continue at the same rate for the whole time; but as the labourers to be employed will probably be such as can be promiscuously picked up, it will be proper to have two sets to relieve each other."

Reducing these figures to the general standard of mechanical power,

$$\frac{4 \text{ feet} + 627 \cdot 5 \text{ tons} + 2240 \text{ lbs.}}{4 \text{ hours} + 6 \text{ men} + 60 \text{ minutes}} = 3904$$

will represent Mr. Smeaton's value of a good English labourer's power, which he estimates as twice that of ordinary persons "promiscuously picked up."

Mr. Smeaton further states, in the same report, that, "If the employment of twelve men for four hours be thought too much, the work may be done in three hours twenty minutes by two ordinary horses." This is calculating rather closely; but it may be taken to mean that he considers the power of a horse equal to that of six men when they work four hours. The dynamical standard of 33,000, however, may be reckoned equal to the power of ten men; and it has been thought right to place this evidence in detail before the reader, that he may make it practically useful.

Progressive Construction of Cranes.

The Walking Crane.—Goods Crane.—The Shipwright's Crane.—Wharf Cranes.—Foundry Cranes.

In the first construction of machines it is seldom that a complete adaptation and fitness for the intended object is at once attained; they are frequently cumbrous and complicated contrivances, and it is not until practice has clearly shown the relation of the means to the end, that machines become simplified, and divested of superfluous material and useless adjuncts.

At all times, and in all circumstances, this is to some extent the case, but in no instance has it been more so than in that of the first cranes which were erected on quays and wharfs for landing and shipping merchandise.

ON THE CONSTRUCTION OF CRANES. 27

The cranes of the last century, especially in the first half of it, were rude and clumsy devices borrowed from the Dutch, many of them worked by men walking within a large hollow wheel, as the turnspit-dog used to do at the same period. Some of these machines lately were, and probably still may be, remaining on the banks of the Thames; and, in the school days of the author, such cranes were used for unloading ships at the quay of Newcastle-upon-Tyne, of which some idea may be formed from Fig. 9.

Fig. 9.

The wheel was about 15 ft. in diameter, and that part of its axle upon which the rope was wound was about 14 in.: the rope then passed over guide rollers to the jib of the crane, which projected over the hatchway of the ship and turned upon a pivot, so that it could move round about three-fourths of a circle, and so deliver the goods upon the quay.

In order to lower the goods the men walked backward; but as it sometimes happened that they were overbalanced by the descending weight, a bar or pole of wood was suspended from the axle, so that in such case they might lay hold of it, and save themselves from being whirled round in the wheel.

The great wheel and the framing which supported it were contained in a wooden building, or rather the beams of the framing covered with weather boarding formed the house, and served also to support the jib, which was attached to one corner of the house.

The first improvement on these primitive cranes seems to have been the liberation of the men from the wheel, which, being reduced in size, was fixed upon the jib; the

Fig. 10.

jib being produced behind the upright to receive and carry it.

The upright or crane-post was fixed in the ground like a mast, with a pivot on the top of it upon which the jib turned. This crane is still in use, and as it may be employed with advantage in the colonies and in new settlements, where timber may readily be had, but where foundries have not been established, it may be well to illustrate the description by a sketch.—(Fig. 10).

This crane is almost entirely of wood, with a small quantity of smith's work, easily forged and fitted, the only part of the iron requiring skill to forge being the crane-handle, in place of which a wooden cross may be used.

The second advance seems to have been the shipwright's crane, fitted with a wheel and pinion; it still retains its original shape, and is an excellent machine for its peculiar purposes, the landing and shifting of timber, and the hoisting of the various pieces to form the frame of the ship whilst building.

The jib is long and lofty, and is firmly secured to the upright or crane-post by a strap of iron on the back, and supported in front by an oak-tree, and a stay of timber morticed into the post, and extending nearly to the point of the jib; a frame behind the post carries the barrel and wheel-work. The post, generally of oak, is placed in a well, and turns upon a pivot at the foot, and in a collar of timbers at the well-top; the collar is lined with iron, and the post is hooped at this part to prevent abrasion and lessen friction.

Although the general form of this crane remains unaltered, yet, in many cases, posts, first made of wood, have been gradually superseded by cast-iron work. The knee, the collar, and the crane-post are, one or more of them, now frequently made of cast iron, and these last-named parts are often bored and turned in the lathe, or fitted with anti-friction wheels, so that the post may more easily revolve in the collar.

30   ON THE CONSTRUCTION OF CRANES

Fig. 11. FIVE-TON CRANE.—Side Elevation.

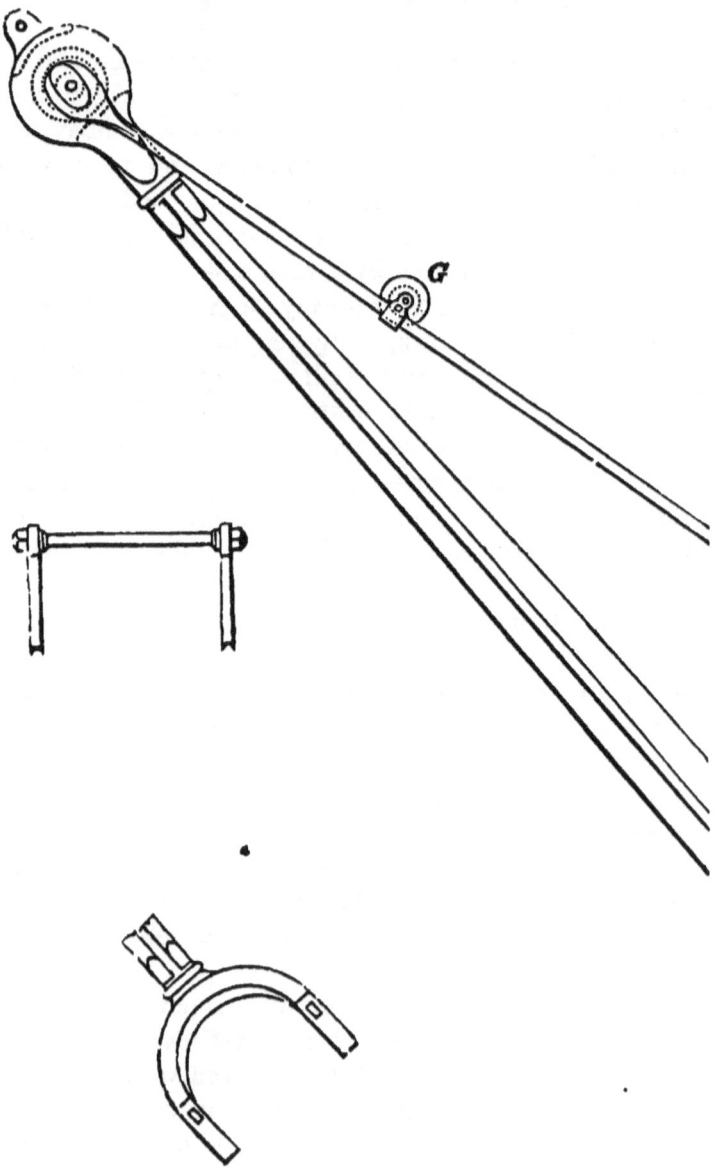

Fig. 11. FIVE-TON CRANE.—SIDE ELEVATION.

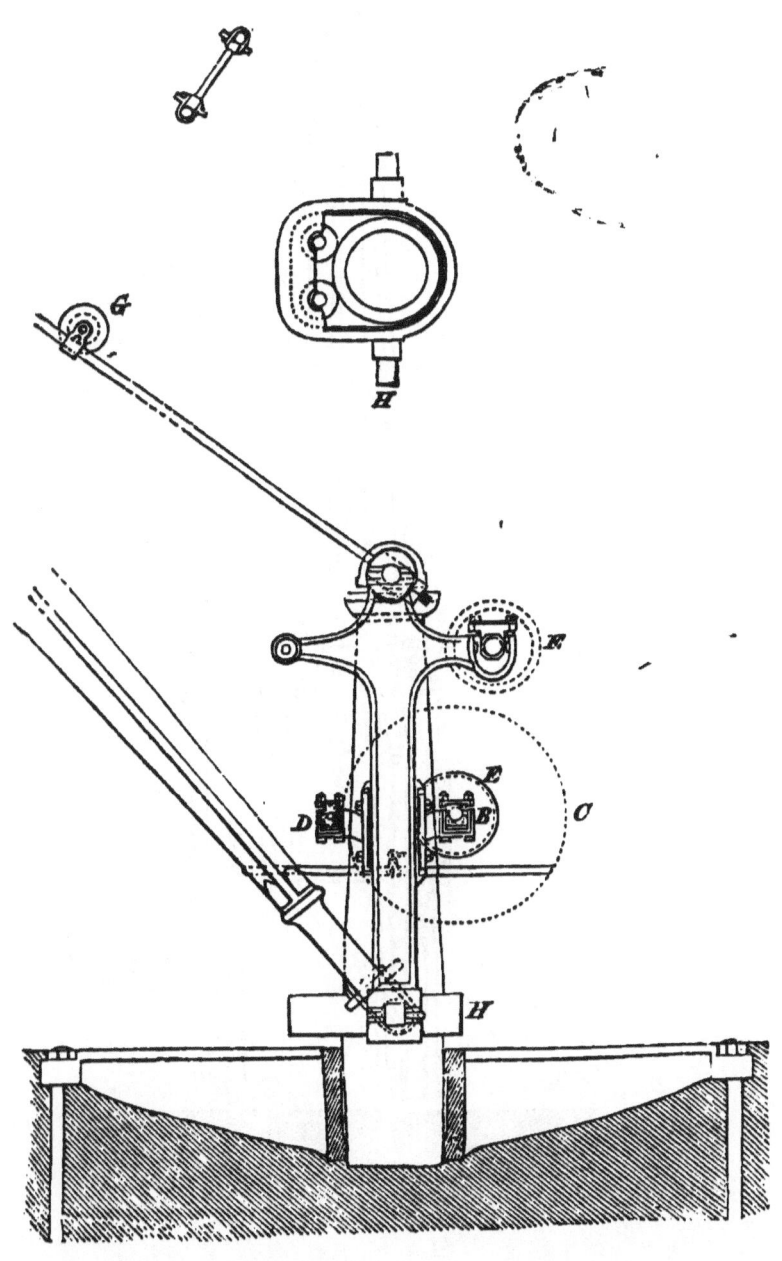

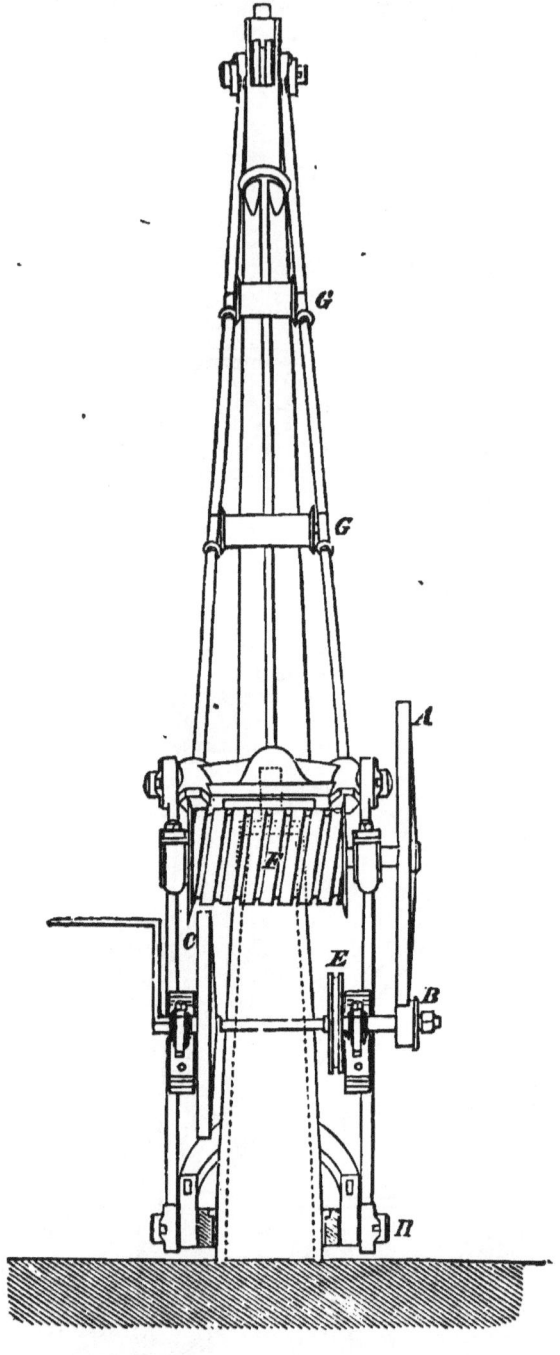

Fig. 11 *a*. FIVE-TON CRANE.—Back View.

The introduction of cast iron for the posts of cranes led to other alterations in their structure; the jib has sometimes been brought down from the top of the crane-post to the foot of it, and the stays of timber formerly below the jib have been replaced by wrought iron "tension bars" above it, extending from the top of the cast-iron post or frame-work to the outer end of the wooden jib, which then acts on the thrust; so that a crane thus made is composed of three different materials—cast iron, wrought iron, and wood. Sometimes the jib also was made of cast iron, as shown in the 5 ton crane. (See Figs. 11, and 11 *a*.)

"Brakes," or as the word is sometimes written, "breaks," have been applied to facilitate the lowering of the goods. These are levers which bring into close contact with a plain wheel, generally fixed on the barrel of the crane, a segment of tough wood strengthened by an iron strap, which by its friction prevents the weight from accelerating as it is lowered.

Thus it is evident that to construct a crane properly a knowledge of the strength and application of materials is necessary; and to calculate the stress to which they are subject, some acquaintance with the composition and resolution of forces is also requisite; and to proportion the power applied to the due performance of its work, it is needful that the mechanical powers should be studied and learned, so that the greatest mechanical effect may be obtained; and also that the construction of wheels and pinions in all their parts, especially that of their teeth, should be well and carefully considered.

Until all these things have been learned, and not before, the making of a crane, simple as the machine may seem, will be little better than guess work; and when cranes are intended to raise such heavy weights as they have now frequently to sustain, their construction should never be entrusted to ignorant and unskilful men, whose mistakes may endanger both property and human life.

ON THE CONSTRUCTION OF CRANES.

Fig. 12. TEN-TON CRANE.

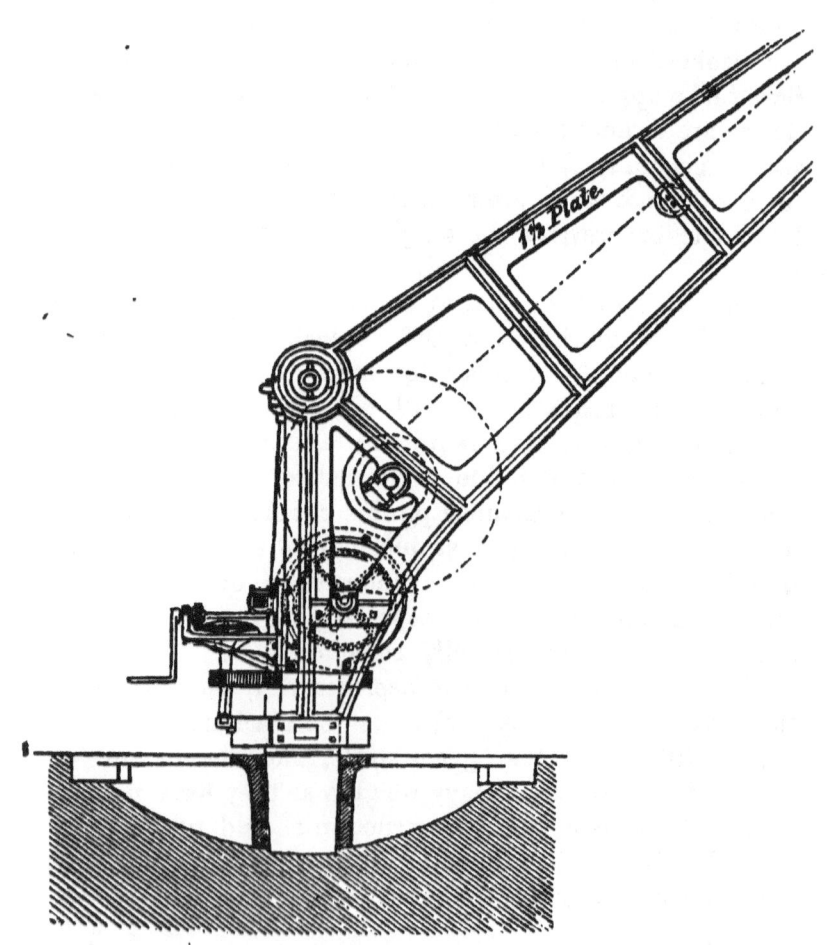

ON THE CONSTRUCTION OF CRANES. 35

Fig. 12. TEN-TON CRANE.

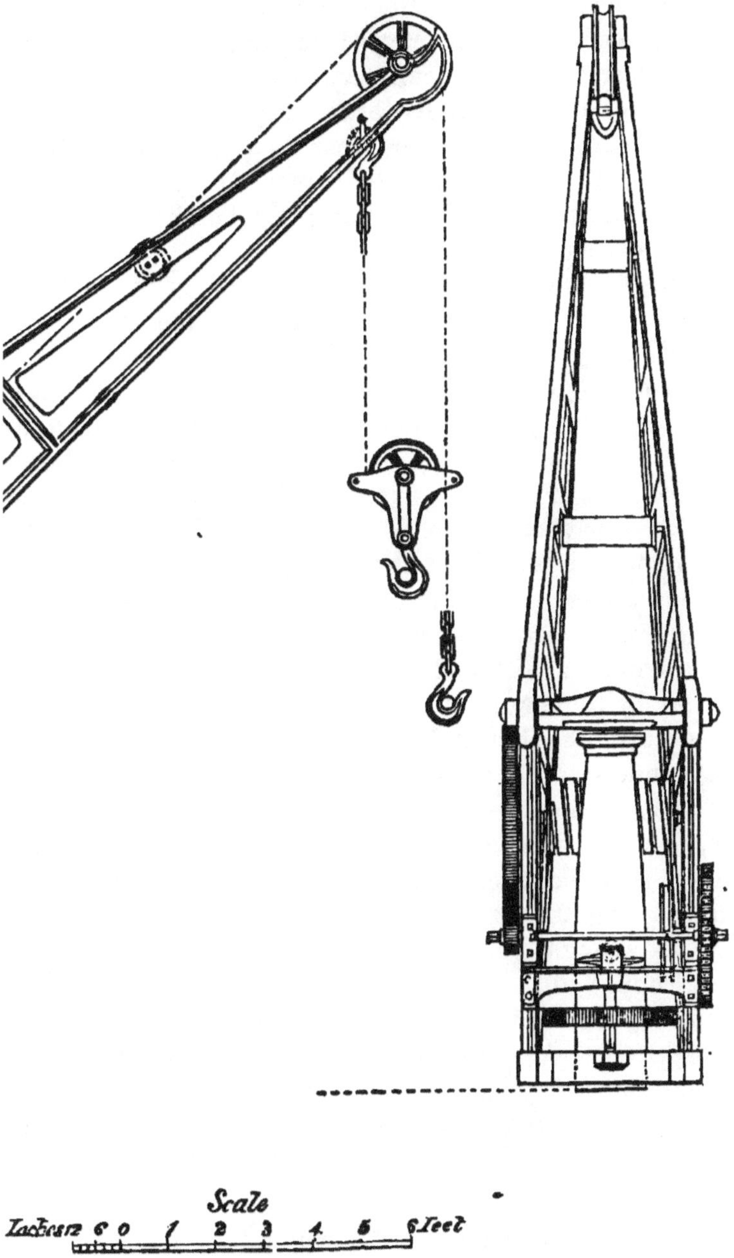

It is not possible in this brief treatise to explain these subjects fully, but they cannot altogether be passed over, and the reader will be referred to those works in which farther information is clearly and explicitly given.* And here it may be remarked, that brakes ought to be applied and used with great caution; and that, as a general rule, they ought not to be attached to cranes for lowering loads greater than ordinary merchandise, or building stones of average size and weight.

Cranes fixed in a well or pit may take a great variety of shapes in their superstructure; sometimes the jib may be long and lofty, as in the shipwright's yard; sometimes it may form a right angle with the post. In some cases it is supported by stays, or struts, in others it is sustained by tension rods; sometimes it is made of timber, and sometimes of cast iron, the form changing with the purpose to which the crane is to be applied; and it is in this adaptation that the skill of the constructor is displayed.

The well-crane having been found inconvenient for raising great weights, because of the insufficient resistance of the ground at the well top, which needed to be strongly secured by framework or by masonry to sustain the pressure against the collar, and the use of cast iron having become better understood, another change of construction took place

The crane-post became a strong hollow pillar of cast-iron, stepped into a massy cross of the same material, bedded in a block of masonry, and held down by strong bolts passing through the mass of masonry to its foundation. A wrought-iron pivot, steeled and hardened upon the point, supported the superstructure, also of cast iron, which turned upon this pivot, revolving round the post, which remained fixed. A cap of steel or of bell metal prevented the abrasion of the pivot; and machinery was in many cases applied to make the crane turn round the post. In cranes of this kind, the stress borne by the post immediately above the cast-iron cross is very great; the action is

* Papers of the Royal Engineers in 4to, Vol. 4.

like the claws of a hammer applied to draw a nail, the weight acting at the jib end as upon a lever, tends to break the post at the level of the ground, or to overturn the mass of masonry in which it is fixed.

An adequate degree of strength is therefore requisite, and in order to obtain it with the least expenditure of material, the post is made hollow. The works of creation, which alone are perfect, teach us by variour examples, as in the stalks of corn, and the feathers and bones of birds and other animals, that a hollow cylinder, or prism, is much stronger than one made solid with the same quantity of material; and also, that if the hollow beam have the hollow or pipe not in the middle, but nearest to that side where the fracture is to end, it will be so much the stronger, as we may see in the wings of sea-fowl, and other birds of rapid flight.

This principle has been adopted and initiated in one of the boldest works of modern times, the hollow railway bridges over the Menai Straits and Conway River, designed by Mr. Robert Stephenson.

The strength of a solid cylinder to resist lateral stress, is as the cube of its diameter; but if the cylinder be hollow, its strength is represented by the difference between the cubes of its external and internal diameters. For example, in a crane intended to carry a load of 10 tons, of which engravings are given, See Fig. 12, the crane-post is made 18 inches diameter outside, and $3\frac{1}{2}$ inches thick, so that its internal diameter is 11 inches

The cube of 18 being 5832, and the cube of $11 = 1331$; their difference, 4501, represents the proportional strength of the hollow crane-post.

A solid post of equal strength must be made full $16\frac{1}{2}$ inches in diameter, for the cube of 16·5 is 4492·126; but the section of the hollow crane-post contains only 160 square inches of iron, whilst the solid post contains nearly 214, making a difference of 54 inches, or about one-third more metal than there is in the hollow post, although it is equally as strong as the solid one.

B *Fig.* 13. FIFTEEN-TON CRANE.

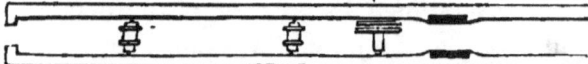

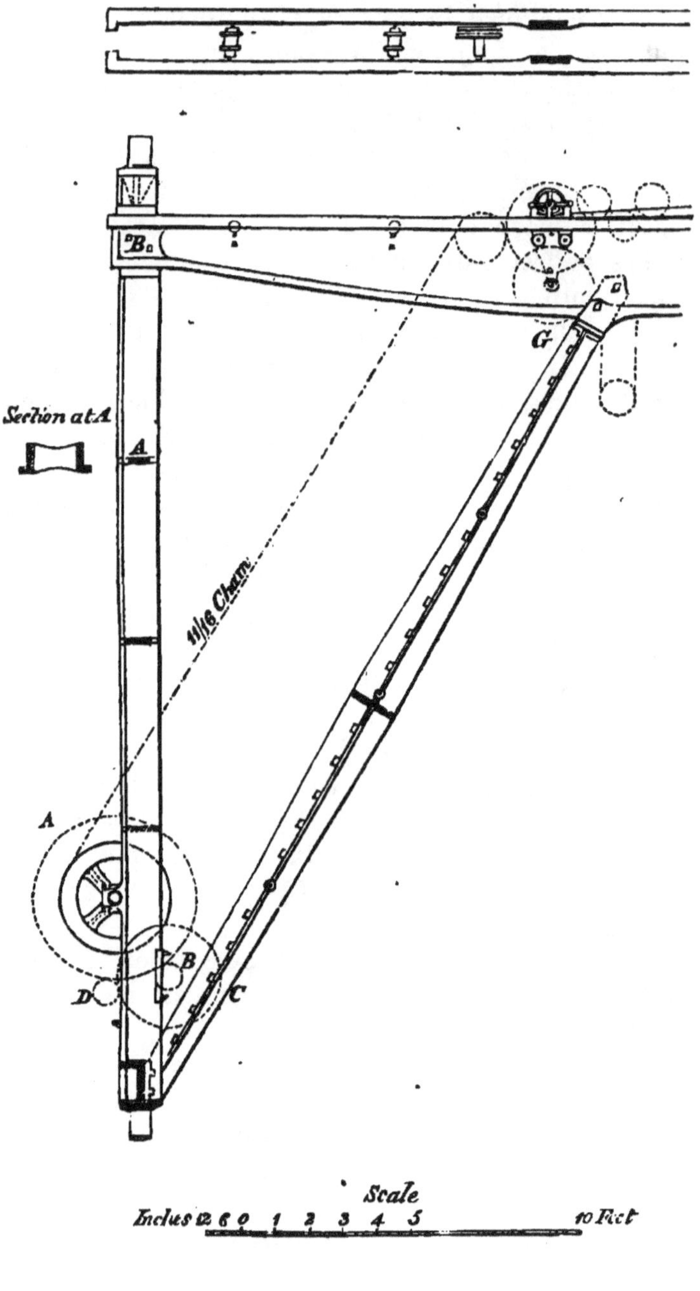

Fig. 13. FIFTEEN-TON CRANE.

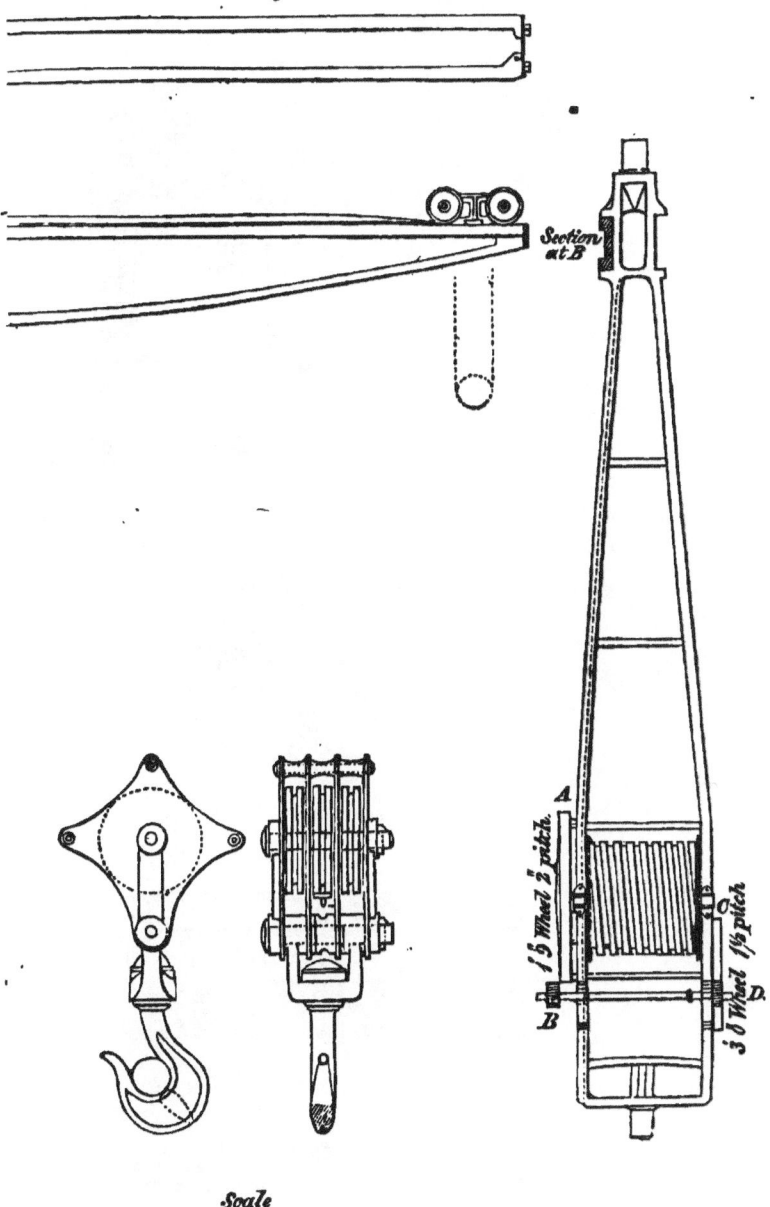

40   ON THE CONSTRUCTION OF CRANES.

ENLARGED SCALE OF DETAILS.

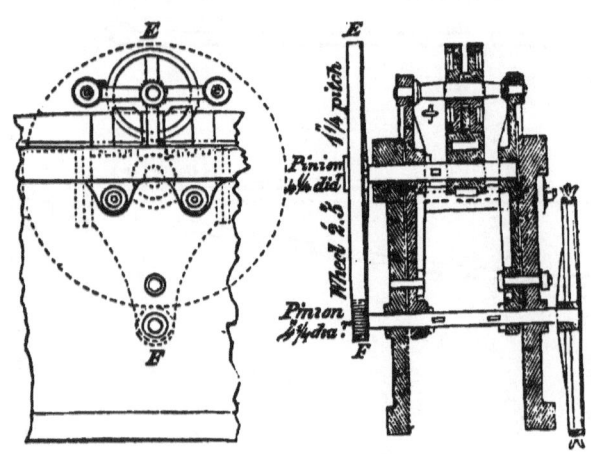

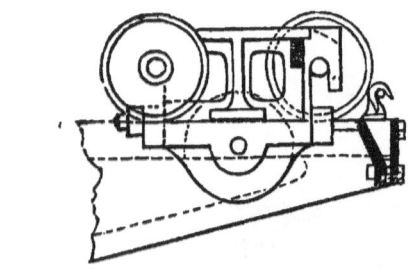

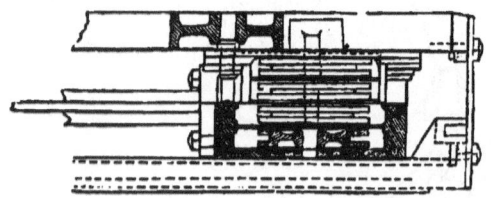

When marine steam-engines had so much increased in size and power that it became necessary to construct cranes specially to deliver and to land their machinery and boilers —cranes calculated to lift weights of 20 or 30 tons—it was no longer safe or practicable to throw so great a stress on the post; it was, therefore, used merely as the centre of a strong circular track of iron, bended upon solid masonry, sometimes built of granite, upon which a massive carriage or framework of cast iron, containing the wheel-work, revolved.

To this framework strong tension bars are attached, and the jib acts as a "strut," resisting the principal part of the stress upon the thrust; wheels, which move upon the circular track, are used to diminish the friction, and machinery is applied to turn the crane round about the central pivot.

Such cranes as these are obviously few in number, their use being limited to the larger seaports frequented by steam ships, most of which have not more than one or two of them. Probably there are not two exactly alike; and as considerable skill is required to make such powerful machines with economy of materials and labour, no opportunity should be lost by the young student or practitioner in examining such examples as may present themselves. In each of them he may find something to learn and something to avoid in his future practice.

The demand for large and heavy cast-iron materials required the erection of powerful cranes in the foundries, and caused them to assume a peculiar form, which may be designated as that of "the foundry crane."

These machines are used not only to lift and move the heavy iron castings, but also to put together and adjust the moulds in which the melted metal will receive its shape. These moulds are sometimes made of dried loam, and sometimes of damp sand, rammed into a frame of cast-iron, which workmen call a flask or box, although it resembles neither of these, but is more like the skeleton cases used for packing glass. Such moulds are both heavy and fragile; the damp sand detaches itself with the slightest jerk, and great accuracy is requisite in putting the parts of the mould together.

In such cranes the upright or post is generally equal in height to the side walls of the building, with pivots at each end, the one turning in a footstep fixed in masonry upon the ground, thn other in a collar secured by framework to the roof and side walls of the building. The jib is horizontal, secured to the post nearly at the top, and at right angles with it. The jib is usually made in two parts, attached to each side of the post, and parallel to each other, and is supported by two stays springing from the front of the post, near to the lower end of it, and fixed to the jib about half way between the end of it and the crane-post.

In order that the crane may work easily and smoothly, the chain must not be large or heavy, as it might be apt to jerk as the links come upon the barrel; therefore, to obtain the requisite strength with a comparatively slender chain, blocks, generally threefold, are used, or rather the lower block has three sheaves or pulleys, and the upper block has four. This fourth pulley permits one end of the chain to be carried to the point of the jib, and fastened there, whilst the other end, after being reeved through the block, passes over a single guide pulley fixed near the middle of the jib, and is rolled upon the barrel of the crane. By this arrangement, the upper block may traverse along half the length of the jib, that is, from the point to the junction of the stays; and by means of traversing machinery, worked by an endless chain descending to the floor of the foundry, the position of ihe upper block, and consequently of the suspended load, may be regulated with the greatest exactness and facility, the load being neither raised nor lowered while the block travels along the jib.

This description will be better understood by referring to the engravings of a crane designed by the author for the foundry of Messrs. Miller, Ravenhill and Co., at Blackwall. See Fig. 13, &c.

TRAVERSING OR TRAVELLING CRANES,

For Stacking Timber or Stone, and for Erecting Buildings, Bridges, and other Engineering Works.

All the cranes hitherto noticed turn upon a centre, and describe a circle with a radius seldom exceeding 25 feet, they can therefore lift only those objects that come within their sweep, and then but such as are within a very short distance of the circumference of the circle they describe, with the exception of the foundry crane, which will lift anything within the area comprised between the two circles drawn by the point of the jib and a radius of half that length, because the upper block travels along the outer half of the horizontal jib.

The fixed centres greatly limit the utility of these cranes, for the load to be lifted must be brought to them, consequently their chief employment is the loading or unlading of heavy goods and materials into ships or waggons, and they cannot distribute their burdens over any extent of surface. It was, therefore, very desirable that the cranes themselves should be capable of moving to certain distances.

One of the earliest examples of the traversing crane was designed by the late Mr. Rennie for the Mahogany Sheds at the West India Docks, where several of them may now be seen at work.

A kind of railway is constructed in the roof, upon parallel frames of timber extending across the building, and upon this a carriage, which is fitted with the wheel-work of a powerful crane, and mounted upon low wheels, travels from side to side of the house. Several of these railways are placed across the shed, so that the largest logs of mahogany can be stacked in rows across the house.

The chain comes down to the floor between the two lines of rails, and the carriage, with the log of mahogany suspended from it, is moved by machinery attached to the carriage, and worked by the men at the crane.

But these cranes have only one motion, that is to say, across the building, and several of them are required for

Fig. 14 a. TRAVERSING CRANE.

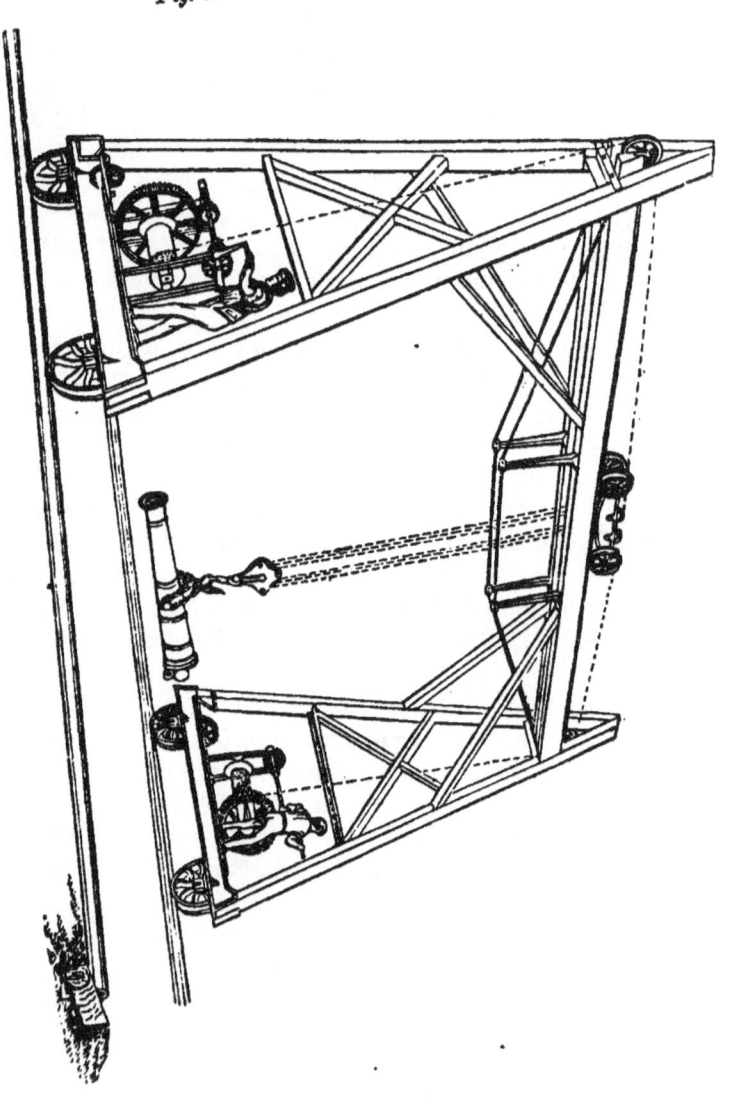

the service of one long shed; so, in course of time, persons began to consider that, if, by some further contrivance, the

Fig. 14 *b.* TRAVERSING CRANE.—END ELEVATION.

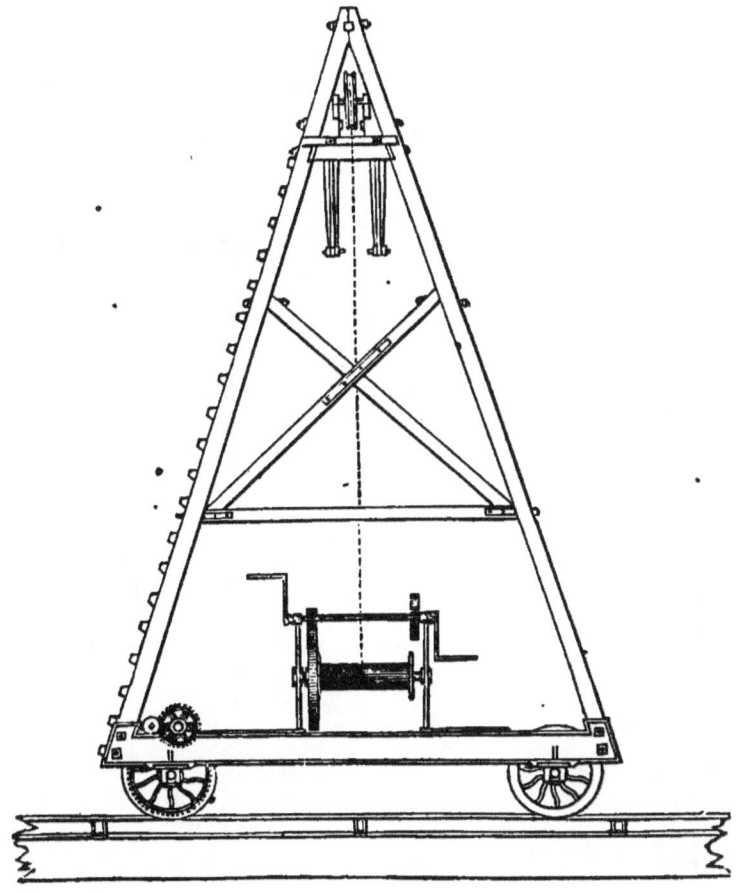

framework which carried the railway could be made to move along the building at the same time the carriage was

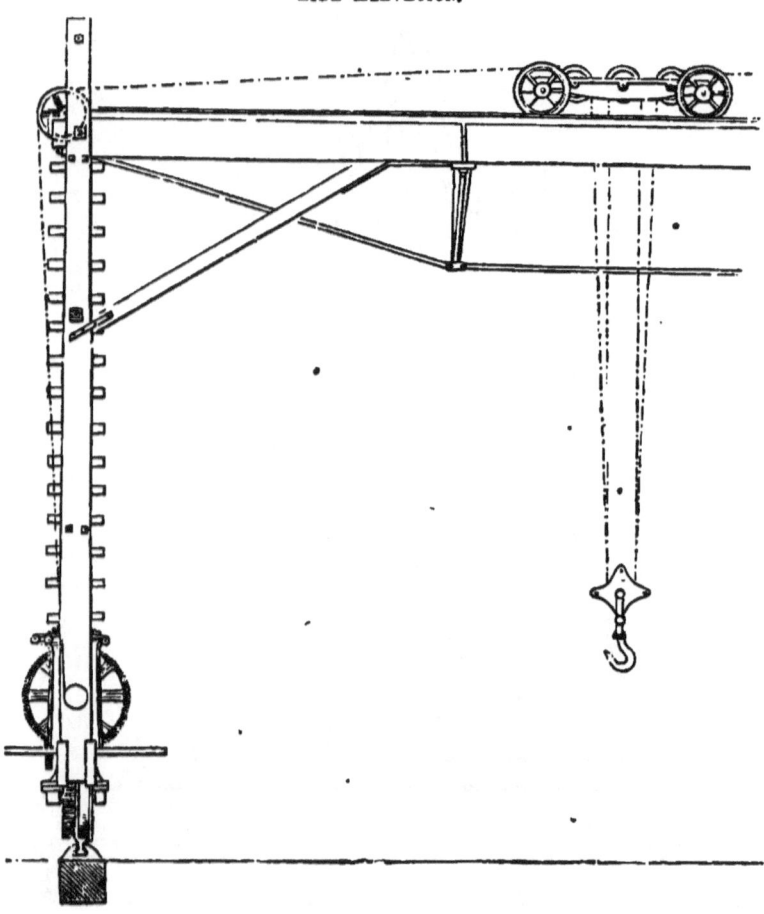

Fig. 14 *c.* TRAVERSING CRANE.
Side Elevation.

travelling across it, they could, by combining the two motions, command the whole area of the floor; and that such

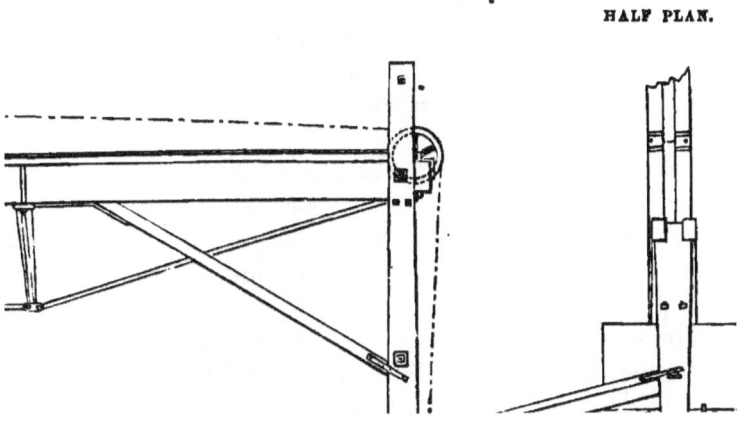

Fig. 14 *d.* TRAVERSING CRANE.

HALF PLAN.

a system of machinery would be most valuable in many cases, but especially in the erection of large marine steam-engines. In designing new buildings for steam-engine manufactories, the side walls are now generally made of sufficient strength to carry a line of rails on each wall, upon an offset in the masonry

On these rails rest two parallel frames of timber, mounted on low wheels at each end, and bolted together at a proper distance apart, so that the frames travel along the building from end to end.

Upon these frames of timber, extending across the building a railway is laid, and a carriage travels as before described, similar to those in the mahogany shed, fitted with crane-work, from which the chain depending may reach the floor at the place desired.

The usefulness of such a combination has caused its introduction in many public works and in several private establishments. Mr. Cubitt, and other eminent builders, perceiving the advantages resulting from it, have employed it extensively, not only in their building contracts, but also in landing and stacking heavy blocks of stone. In these cases the longitudinal railways are carried upon beams of timber resting upon uprights ranged in rows along each side of the stone-yard, or of the building to be erected; in which instance the uprights, being lofty, are secured both by horizontal and diagonal ties. Indeed, so many and so great are the benefits derived from such a system of framing and machinery, that Mr. Grissell thought it right to construct framework the full height of the Nelson Column, in Trafalgar Square, with a travelling crane at the top strong enough to place the statue on the pinnacle. Mr. Grissell has presented a model of this framework to the Institution of Civil Engineers, in order that others may profit by his experience; and similar machinery is also used in building the new Houses of Parliament.

Other cases, however, have arisen in the modern practice

of engineering that render it desirable to dispense with framing, as not only expensive but inconvenient. In the erection of cast-iron bridges, the parts of which are fitted together at the iron works where they are made, it is necessary, both for convenience and despatch, that the hoisting machinery should embrace the whole fabric, and pass along at the same time.

The Butterley Company, in the erection of cast-iron bridges, use cranes of this kind, which are strong, simple, and easily managed, with less machinery than has been generally thought necessary for such operations, the whole of which can be worked from below.

The travelling or traversing crane, shown in the engravings (See figs. 14 a, 14 b, 14 c, 14 d), was used to erect the large cast-iron drawbridge over the river Ouse, at Selby, designed by Messrs. Walker and Burgess, for the Hull and Selby Railway Company.

It is composed of two triangular frames of timber, based on cast iron plates set edgewise, and mounted on wheels similar to those of railway waggons.

These frames support two parallel beams of timber, trussed underneath by wrought-iron tie-bars an inch and a quarter in diameter, and cast-iron struts; on these beams is laid a railway, upon which travels a carriage containing the pulleys for the chain, and constituting a fourfold purchase block, the chain passing between the two beams to the lower blocks. The ends of the chain pass along and above the beams to the fixed pulleys at each end, and thence down to the winches, which are secured upon the cast-iron bases of the two triangles.

By winding one of these winches, and unwinding the other at the same time, the carriage and the load suspended from it travel from one end of the beams to the other; and by winding or unwinding one of the winches only, the load is raised or lowered.

The waggon-wheels, on which the triangular end frames are mounted, have toothed wheels and pinions attached to them; by turning the pinion handles the waggon-wheels are made to revolve, and the whole fabric, with its load, travels like a locomotive engine along the two parallel lines of railway on which it is placed; these rails are of the strong kind used on the London and Birmingham Railway, and are laid 30 feet apart. This transverse distance may be called the span of the crane; the longitudinal distance is, of course, limited only by the length of the railway on which the crane travels.

The crane here represented will lift a weight of eight tons, and the cost of it, including the chain and blocks, was 150*l*. It is readily taken to pieces for removal; and, by the combination of its movements, every point within the area comprised between the rails may be commanded with the greatest exactness and facility. Hence the utility of this crane for the purpose of fitting together the heavy portions of large work, which cannot be done by a crane moving round a fixed centre. It may also be made very useful in storing and stacking heavy materials, as stone, timber, anchors, or cannon. The traversing crane is used with great advantage in working the diving bell, when it is employed for laying the foundations of sea-walls under water, as the bell by this means may be made to travel along the line of wall; and the stones may be lowered so nearly into their places, that they require but little adjustment by the divers.

Self-acting Hoisting Machines.

The Sack Tackle.—The Lift.—The Cornish Man-Machine.—The Vacuum Crane, Steam Crane, Water Crane, &c.

The successive improvements of the crane, already detailed, have not been sufficient to satisfy the minds of ingenious mechanics, and many attempts have been made to render cranes self-acting; so that manual labour might be dispensed with, and intelligence alone be required to direct their operations

Some of these attempts have been attended with considerable success. Skilful millwrights, engaged in the construction of extensive corn mills, have, in many instances, made most ingenious mechanism for transferring the grain and the flour to various parts of the mill; and hoisting tackle for raising the wheat and delivering the meal in sacks, worked by the machinery of the mill, has been used for a long time past. The machinery of the sack tackle is called into action by pulling a cord held in the hand of the miller; and it continues in motion until he slackens his hold, whilst the heavy sack of corn, no longer carried on his shoulders, but becoming as it were obedient to his will, rises through successive trap-doors and lays itself at his feet.

The lift or hoist is also used in cotton mills to raise a kind of ascending room, in which not only the materials, but the workpeople themselves, are carried from floor to floor; and the waste of their physical strength, otherwise expended in repeatedly climbing the stairs, is avoided.

Persons rising from the ground floor in the ascending chamber, are landed on any of the upper floors they please, by disengaging the machinery of the hoisting tackle, which much resembles that of the corn-mill.

The waste of strength arising from the exertion of climbing a long flight of stairs or a series of ladders is very great, so much so that in the Cornish mines, where the miners had to ascend crooked shafts by means of ladders from their daily labour, their strength was often so far spent, when they reached the top, that they were fain to lay themselves exhausted on the ground. The ascent from the mine was the most severe portion of the day's work. It was difficult in many of the mines, when the shafts were neither straight nor perpendicular, to find a remedy for such unprofitable toil, as ropes could not be applied in such shafts.

The mining companies, therefore, offered a premium to the inventor of the best machine for "bringing the men to grass." Many ingenious plans were proposed, but the best was one in which two rods of timber move with a pumping action parallel to each other, and only so far asunder that a man may step without danger between the two rods, working like pump rods, with a stroke of about 12 or 14 ft., and an alternating motion, one rod moving upward as the other moves downward. They are fitted with suitable guides and rollers to keep them steady in work, and to prevent needless friction, and each rod makes 3 or 4 strokes per minute.

Upon each of these timbers is fixed, at regular distances equal to half the length of stroke, whatever that may be, say 6 or 7 ft. apart, a series of steps or small projecting scaffolds, whereon one man may stand, and there are also long staples or holdfasts which he may grasp with his hands.

The rods are almost stationary for a second of time, when the crank which moves them is turning past its centres, and at this period, which is the termination of a stroke, the stages upon the two rods coincide with each other; a man, therefore, can step upon a stage of the ascending rod, and he is carried up 12 or 14 ft.; he passes over to the other rod which has just descended, and finds the stage upon it ready to receive him and carry him up another stroke; thus, by stepping alternately from rod to rod, he rises with a zigzag movement, and with very little fatigue, in a short time to the mouth of the pit, travelling about 250 fathoms in 20 minutes.

As soon as he has quitted the first stage of ascent another miner takes his place, and thus a body of men continues rising from the mine, the rods being loaded with people for their entire length, until the whole party has ascended, the relief party descending at the same time on the alternate steps.

A model of this man-machine may be seen at the Museum

of Practical Geology, an institution that ought to have more numerous and frequent visitors. This machine is worked by steam power, and is mentioned in this place somewhat out of its regular order, as it is a recent invention.

The first successful attempt to work detached cranes by mechanical power, and without manual labour, appears to have been that of Mr. Hague of London, who, by means of an air-pump worked by steam power, exhausted the air from a receiver, and continued by pumping to maintain a vacuum in it. From this receiver pipes were laid under ground to several cranes in one of the docks, and each of these cranes was fitted with a small cylinder, vibrating or oscillating upon hollow pivots behind the crane-post, with the piston-rod acting upon the pinion shaft of the crane's machinery, which was fitted with a crank and fly-wheel for that object.

The cylinder resembled that of an oscillating high pressure steam-engine, except that its power was derived from the exhaustion of the air and not from the pressure of steam.

There are, however, several objections to machinery worked by means of an exhausted receiver; it is very liable to derangement from the leakage of joints and fittings, and difficult to manage and adjust in the hands of ordinary workmen. The vacuum cranes therefore went out of use, although similar machinery for winding coals is still employed underground in pits where the presence of inflammable air renders it unsafe to place the fires of a steam-engine boiler, the vacuum pipes being carried down the pit.

While the vacuum apparatus remained in abeyance, it occurred to another ingenious mechanist that if high-pressure steam were applied directly in the cylinder attached to the crane, he might dispense with the air-pump, the receiver, and their adjuncts. This idea was carried into effect, and cranes, with small steam-engines fixed to them,

54 ON THE CONSTRUCTION OF CRANES.

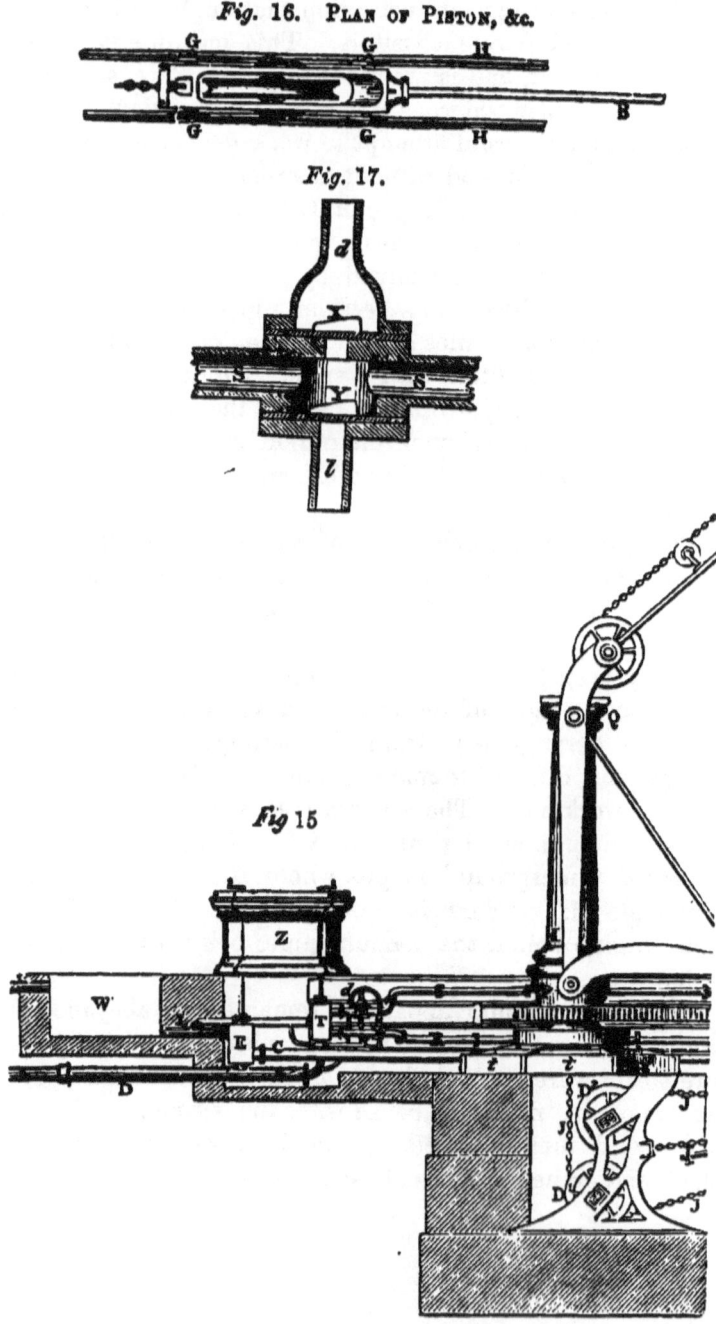

Fig. 16. PLAN OF PISTON, &c.

Fig. 17.

Fig 15

ON THE CONSTRUCTION OF CRANES.

Fig. 15.—WATER PRESSURE WHARF CRANES.

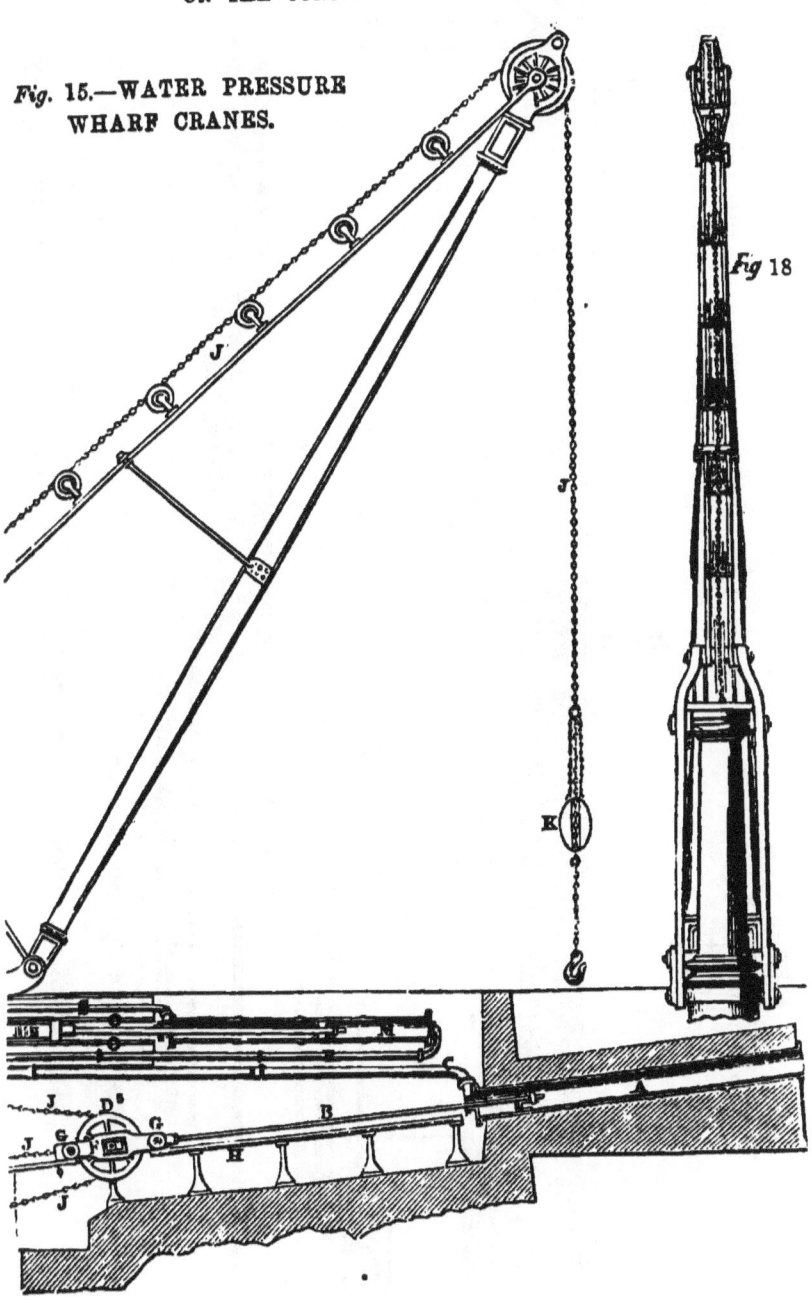

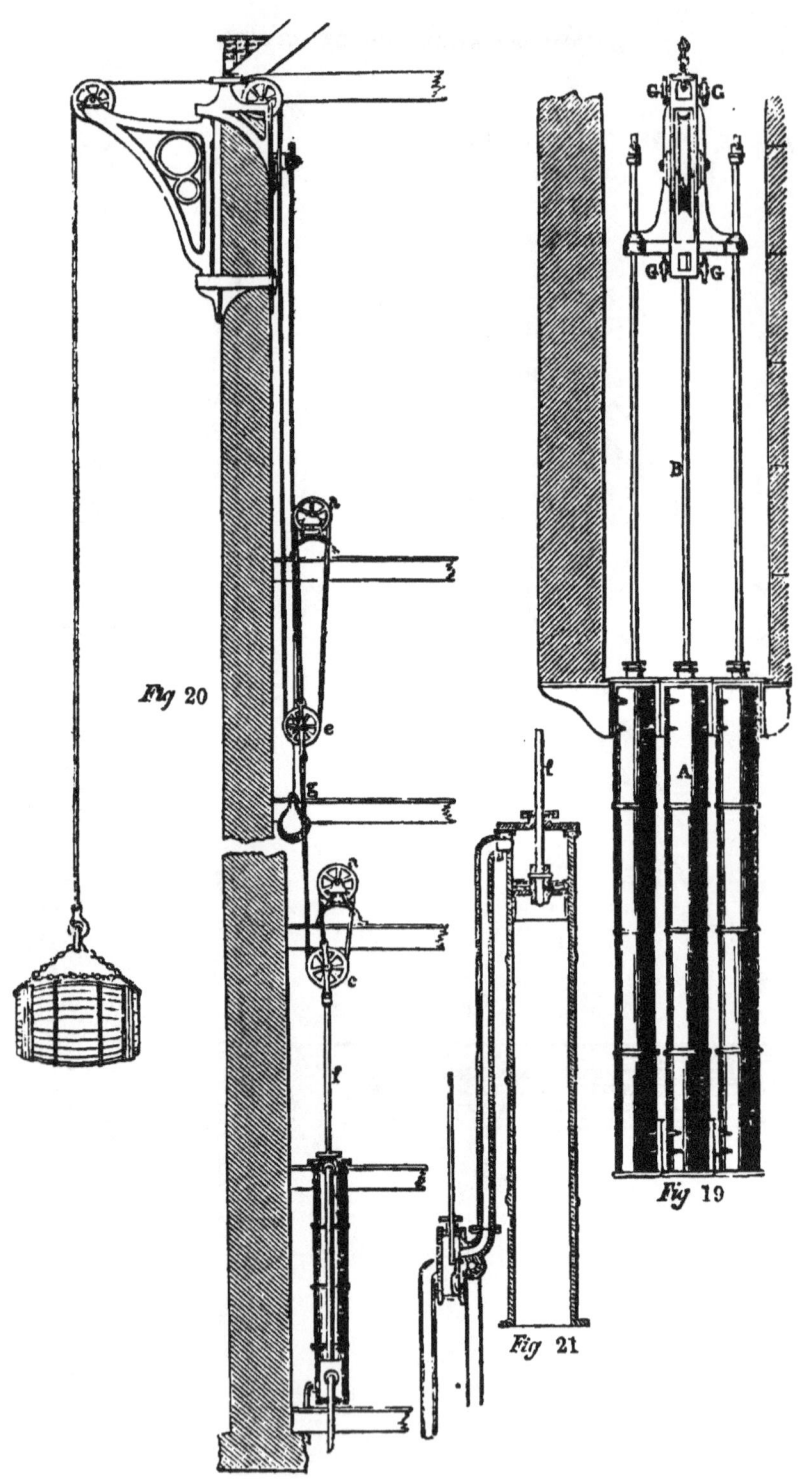

ON THE CONSTRUCTION OF CRANES

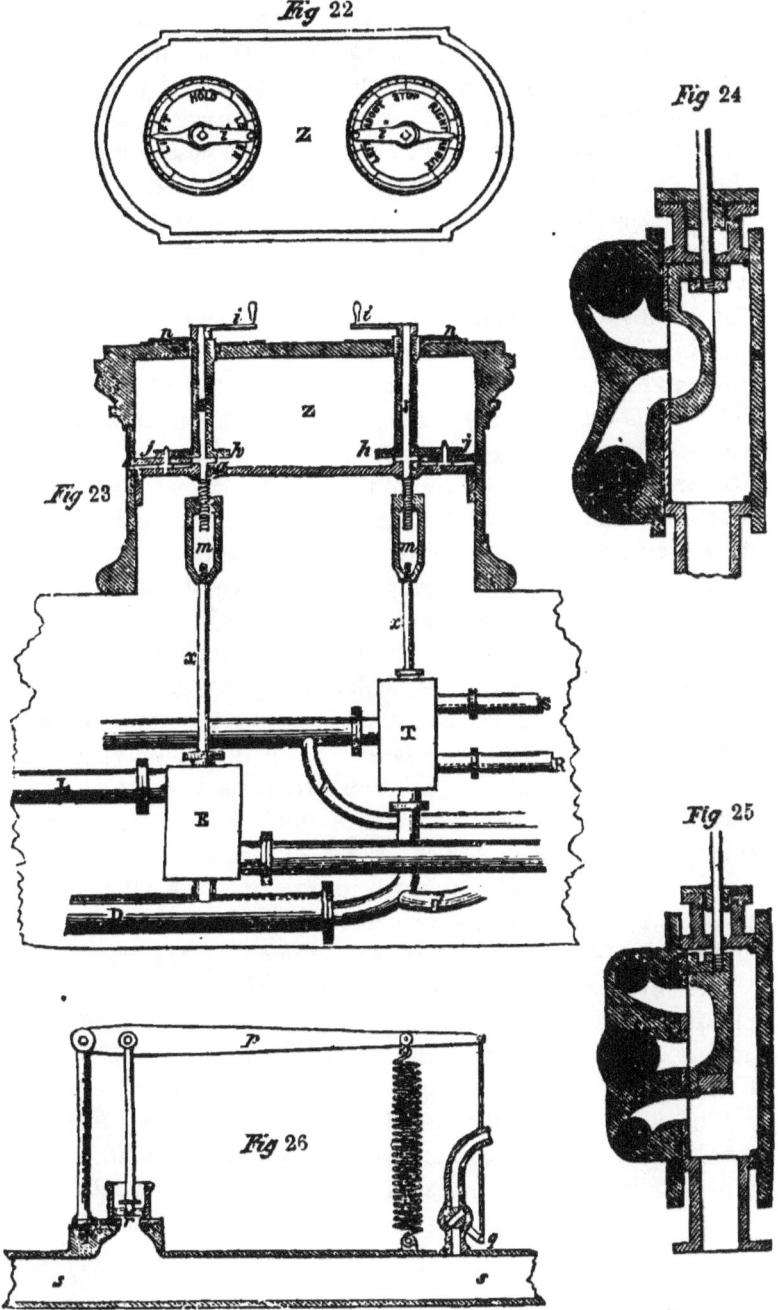

are worked with advantage in cases where loads varying but little in weight require to be hoisted in continual succession. The author had lately an opportunity of seeing one of these cranes in action on the new works at Dover Harbour, hoisting blocks of stone from the coasting vessels. The crane was fixed upon the quay, the boiler was in a building not far distant, the steam pipe was conveyed under the road, and then rising up was connected to a swivel joint on the centre of the crane-post, and conducted the steam to the oscillating cylinder suspended by hollow pivots above the pinion axle; this was cranked to receive the cap of the piston-rod, and fitted with a small fly-wheel. The crane was skilfully managed by a workman, who, upon the ascent of the stone, stopped the little engine, applied the brake, disengaged the pinion, and lowered the stone upon a truck, in about the same time that has been taken to write this brief description of his movements.

A self-acting crane differing entirely from these has been erected on the quay at Newcastle-on-Tyne, from a design by Mr. Armstrong of that town. It is worked by water-pressure. The reservoirs which supply the town with water are placed more than 100 ft. above the level of the quay. Three cylinders placed under ground, lying on their sides, and parallel to each other, have their piston-rods attached to one cross-bar, which carries a strong pulley, over which passes the bight or double of the crane chain, one end of which is made fast, and the other end, passing upward through the centre of the hollow crane-post and over guide pulleys, is carried to the end of the crane jib.

The cylinders have a long stroke, and are fitted with pistons and slide-valves, somewhat like those of a steam-engine, but the pulley in the double of the chain makes the hoist equal to twice the length the pistons travel.

The column of water acts by its pressure upon the pistons, and for lifting ordinary weights it is admitted into the centre cylinder only; for extra weights, it is turned into the two external cylinders; and for the heaviest weights, into all the three.

A fourth cylinder, having the piston-rod fitted with a rack acting on a strong toothed-wheel upon the crane-post, serves to turn it round; and the crane chain being weighted just above the hook, overhauls itself without farther assistance.

A case or chest of cast-iron neatly formed, like a pedestal, is secured to some masonry at a short distance from the crane, and contains the apparatus for turning the water on or off; beside this stands the workman, who, by the management of a few sliding stop-cocks, hoists, lowers, and turns the crane at his pleasure.

Cranes of this kind, both for wharfs and warehouses, have been erected at Liverpool, with complete success, at the Albert Docks; a crane worked by a column of water 210 ft. in height has been set to work at Glasgow; it will lift 15 tons, and it is worked with great facility and promptitude Engravings of two cranes are here given; one is for a wharf, the other for a warehouse. The contrivance for using water at so high a pressure requires a more detailed description, which, by the liberality of Mr. Armstrong, the author is enabled to give.

Referring to the Illustrations—*Fig.* 15 shows a side view or elevation of a crane. A is a cylinder containing a water-tight piston, attached to the piston-rod B; C C is a feed pipe which communicates with the main supply pipe D, and by which water under pressure is caused to enter the cylinder at pleasure, by means of the slide-valve E; D^1 and D^2 are fixed pulleys, and D^3 a movable pulley which travels with the carriage F attached to the end of the piston-rod, and supported on the friction rollers G G, two of which are placed on each side of the carriage, and run upon rails H H, as shown also in the Plan (Fig. 16). I is a hollow cast-iron pillar, forming a fixed centre, round which the movable parts of the crane turn; and $t\,t$

is a cast-iron frame in which the pillar is stepped, the frame being bolted to the foundation.

J J J J is the chain by which the load is suspended, and this chain passes downwards through the centre of the hollow pillar; and after passing over the pulleys D^1, D^2, and D^3, it is fixed to one end of the movable carriage F.

When water from a sufficiently elevated source is admitted into the cylinder, the piston-rod will be put in motion by the pressure of the water, and lift the load attached to the end of the chain; while, the extent of the piston's action being multiplied threefold by means of the pulleys, the load is lifted to a height equal to three times the space passed through by the piston.

And when the water is allowed to escape from the cylinder through the pipe c c, by the opening of the passage in the valve box E, communicating with the waste pipe L, and the closing of that communicating with the main pipe D, the piston returns towards its former position, and the load is lowered.

The cylinder is placed in an inclined position, in order that the weight of the movable carriage, with its pulleys and appendages, may facilitate the overhauling of the chain by the counterweight K.

The crane also turns round in either direction by the action of water, by means of the following apparatus; M is a cylinder containing a piston, acted upon by pressure on both its sides, connected with the rack N, which travels between guides o o.

The teeth of this rack work into corresponding teeth surrounding the lower margin of the collar P, to which collar, and to the cap Q, the framework of the jib is fixed; R R and s s are the pipes by which the water under pressure is conveyed to or from this cylinder at either end, and the admission and emission of the water are regulated by a box slide valve T. Near to T are two valves which may be called the relief valves, and which are fixed, one upon each of these pipes.

The object of these relief valves is to prevent the circular motion of the jib from being too suddenly arrested by the closing of the slide-valve; they will be better understood by referencr to *Fig.* 17, which is a sectional view of one of these relief valves u, showing a portion of one of the pipes s, upon which it is fixed; *d* is a small tube communicating with the main supply pipe, and *l* is also a small tube communicating with the waste pipe L and a small cistern w, which is kept charged by the waste water from the cylinders A and M of Fig. 15.

In these relief valves there are clacks, each opening upwards, marked x and y. The manner in which these relief valves act is as follows:—

Suppose the jib of the crane to be swinging round by the action of the water upon the piston, the pipe s acting at the time as the egress passage; then, on the sudden closing of the slide-valve, the egress being abruptly stopped, the whole of the turning apparatus would be subjected to a violent shock from the momentum of the jib, were it not for the clack x, which there gives vent to the confined water, and allows it to return into the main supply pipe as soon as the compression caused by the momentum of the jib becomes sufficient to raise the valve, by overcoming the weight of water upon it; thus the piston, instead of being suddenly stopped, is only powerfully retarded, and the jib is brought to rest quickly but without shock.

If the pipe s be acting as the ingress instead of the egress passage, then on closing the slide valve, and while the momentum of the jib continues to impart motion to the piston, water would be sucked up through the clack y, from the waste water cistern w, to supply the void which would otherwise be left on the influx side of the piston.

Thus, by applying relief valves to each of the waterways, the detrimental effects which might otherwise result from the swing of the jib are prevented, and the cylinder which turns it is kept always fully charged with water on both sides of the piston, and ready to receive immediate impulse in either direction as soon as the slide valve shall be reopened.

Reverting to *Fig.* 15, z is the index table; the internal arrangements of it are such that the workman, by turning the indices with a cranked handle upon each of them, has the entire control of all the operations of the crane, the index on the right directing its revolving movements, and that on the left the hoisting and lowering.

Fig. 18 is an elevation of the jib and pillar of the crane viewed from behind.

The cylinders, pipes, and valve boxes are underground, secured to masonry, and the excavation is covered with boards, flag stones, or chequered plates of cast-iron.

In the crane at Newcastle, before described, there are three cylinders, and the length of strike in them is 12 feet, which, by the movable pulleys, is multiplied threefold, making the lift 36 feet. The column of water is about 240 feet high.

Fig. 19 shows an arrangement of three cylinders, for the purpose of obtaining various degrees of power.

When the lowest power only is wanted, the water is admitted into the middle cylinder; and, while the travelling pulley is drawn along by the action of the middle piston-rod B, the two outer piston-rods, passing through holes in the cross head, remain at rest, the cross head sliding upon them. When the second power is required, the water is admitted into the two outer cylinders and shut off from the middle one, which then exerts no power; but the piston-rods of the outer cylinders, having stops upon their ends, then act upon the cross head. And when the greatest power is needed, water is admitted into all the three cylinders at the same time.

The power may also be varied by varying the number of pulleys, if they be so arranged.

Fig. 20 represents a combination of pulleys in cases

where very high lifts are required, as in lofty warehouses, and where it is necessary to multiply the comparatively short stroke of a piston.

Fig. 21 is a sectional view of the vertical cylinder shown in the last figure, drawn to a larger scale.

Fig. 22 is the top of the index table, and *Fig.* 23 the remainder of the apparatus connected with it, drawn to a larger scale, and shown partly in section, in order to show how motion is given to the slide-valves.

For this purpose the cranked handle i is fixed upon a rod b, terminating at the lower end in a screw, which works in the hollow head m of the valve-rod x.

The pointer n is fixed upon a tube which turns freely round the rod b, and has a cogged wheel h upon its lower end. The rod b has also a small cogged wheel or pinion g fixed upon it, which works with a larger cogged wheel k, to which is fixed another pinion j, which works with the larger cogged wheel h attached to the tube, which forms the axis of the pointer, thus giving it a reduced motion, so that the pointer makes but one revolution, while the cranked handle makes as many as are necessary to open and shut the slide valve by moving the full length of its traverse.

Fig. 24 shows a section of the slide-valve box E, and *Fig.* 25 a similar view of the slide-valve box T.

Fig. 26 is a kind of safety valve, which may be placed upon the main supply pipe D to prevent its bursting, by the sudden increase of pressure which is liable to take place in the pipe, when the flow of water through it is, from any cause, abruptly stopped. The power of the water, operating on the piston r, overcomes the resistance of a spiral spring attached to the lever p, and opens for a moment the escape-cock q, which relieves the strain upon the pipe.

The application of a column of water to lift weights was made many years ago, by the late Mr. Matthew Murray of Leeds, who employed it to raise the heavy boilers he manufactured for the spinning-mills in that district.

His mode of using it was very simple and effective. From

a cistern placed upon a lofty building, a water pipe communicated with a cylinder set upright upon the top of "a triangle" formed of three stout trees, and fitted with a piston and rod, which passed through a collar of leather or stuffing box in the bottom of the cylinder; on the end of the piston-rod was a loop or shackle, to which strong chains were attached, for suspending large boilers, engine beams, &c.

By admitting the water between the cylinder bottom and the piston, the load was lifted sufficiently high to allow a waggon to pass under it, and, by allowing the water to escape, the weight descended upon the carriage.

Subsequently, Bramah's press has often been used for raising great weights; but in no case has its power been exerted to such effect as in raising the tubular bridge over the river Conway, by Mr. Robert Stephenson.

In a book like this, it may be proper shortly to explain the principles on which the pressure of water acts on the piston of a cylinder, or the ram of a hydrostatic press.

The equal pressure of fluids in all directions, with a force proportionate to the height of their *vertical column*, without any relation to the weight or quantity of the fluid itself, appears at first sight so paradoxical, that it was long known as "the Hydrostatic Paradox," and was illustrated by lecturers on natural philosophy, by means of an apparatus called the Hydrostatic Bellows. (*Fig.* 27.)

Fig. 27.

Two thick flat round boards, of about 18 inches in diameter, were connected by a short hoop or cylinder of pliable leather, firmly nailed, and made water-tight round their edges, and made to open and shut like common bellows, but without valves. A pipe about 4 feet high, communicated with the cavity between the boards, having a funnel at the top, and, when water was poured into the pipe, it gradually separated the boards causing the upper one to rise, even when loaded with a weight of 3 cwt., although the weight of water in a pipe of half an inch in diameter, and 4 feet in height, does not exceed 6 ounces. The pressure of a column of water four feet high acts upon the whole under surface of the upper board, which has an area of 254 square inches. Thus, in Mr. Murray's lifting apparatus, if the cistern were 60 feet high, and the cylinder 40 inches in diameter, the pressure upon the piston would be sufficient to lift a weight of 24 tons.

In Mr. Armstrong's cranes, supposing the hydrostatic column to be 240 feet, which at Newcastle it is, the pressure in his cylinders will be four times as much as that in the cylinders upon Mr. Murray's plan.

Carrying the same principle still further, Mr. Bramah applied to the cylinder of his press a small forcing-pump, with a powerful lever acting on the plunger of the pump, which, in some cases, he made not more than three-fourths of an inch in diameter, so that a man might exert a pressure of a ton or more upon the plunger, and, consequently, a force of two tons upon every square inch of surface within the pump, for the area of such a plunger is somewhat less than half a square inch ($\cdot 4417$), therefore the water forced into the larger cylinder exerted the same pressure upon every square inch of surface within it as has before been shown.

Instead of pistons, Mr. Bramah's cylinders were fitted with plungers, or "rams," working through a collar of leather, the outer end of the ram acting against press plates, or performing any other work requiring heavy pressure, so that if the ram were 10 inches in diameter, a force of 150 tons might be gained by it.

The weight of one of the two tubular beams of iron, forming the bridge over the river Conway, in the line of the Chester and Holyhead Railway, was reckoned at 1300 tons, and it was lifted by two presses, one at each end of the tubular beam; these cylinders were each 20 inches inside diameter, and the rams 18¾ inches, the water space round the ram being full three-fourths of an inch.

The cylinders were about 9 inches in thickness, and the rams had a stroke or lift of 6 feet. On the top of each ram was fitted a massive iron cross-head, from which, passing through holes in it, depended two strong chains, formed of flat wrought iron, in links of eight and nine bars alternately, their length being 6 feet.

By these chains the tubular bridge was raised 6 feet at every lift of the rams, each end of the bridge being suspended by the two chains coming down from the cross head of each press, as before mentioned.

The presses were elevated above the level of the bridge upon strong iron girders, supported on masonry, with openings in the girders for the chains to pass through, and provision was made for securing the chains as they passed through these openings, by means of strong vices or screw clams, so that, each time the tubular beam was lifted 6 feet, the lift was made good, and one link or set of plates removed from the chains. The water was discharged from the cylinders, the rams descended, and another lift was taken, until the bridge was raised to the required height, which was about 24 feet.

The forcing-pumps which injected the water into the cylinders were worked by steam power, at a pressure of 2·14 tons on the circular inch; the section or area of the ram, being 337·64 circular inches, gives a force of 722½ tons to each press, or 1445 tons to the two.

The time required to raise the bridge and build up the masonry to it was four days; but it was actually lifted 13 feet high in an hour—an effort of mechanical power which has never before been equalled—yet each cylinder, when full, contains only 13 cubic feet, or less than 82 gallons of

On the Strength of Materials used in the Construction of Cranes.

Cast Iron—Wrought Iron—and Wood.

It is proposed to conclude this treatise with a brief notice of the strength of materials, and a review of the laws which regulate their proportions and application in the construction of cranes and to other purposes, for raising and supporting weights; but, as such notice must of necessity be very limited and imperfect, references will be made to those works which contain more ample and detailed information for those who may wish to pursue the subject further.

The strength of rectangular beams or bars of wood or metal, supported at both ends and loaded in the middle, is in proportion to the square of the depth, multiplied by the breadth, and divided by the length between the supports.

Hence it is that beams and joists are made deep and narrow; for a beam which is 6 in. broad and 12 in. deep will carry a greater weight when set edgeways than when laid flat, in the proportion $12^2 \times 6 = 864$ to $6^2 \times 12 = 432$, or 2 to 1; this rule holds good so long as the beam can be kept from twisting, and is quite straight in the grain, and free from knots or flaws; consequently, flooring joists which are steadied by the boarding and by cross bracing, are often made 9 in deep by 2 in. thick, or 12 in. deep and 3 in. thick, which proportions answer the purpose very well in such circumstances. But when a single beam of large dimensions, or a piece of strong and independent framework is required, the best proportion of depth to thickness is as the square root of 2 is to 1, or nearly as 7 is to 5. It is also the strongest piece of timber that can be cut out of a round tree; for, although the quantity of timber contained in a square beam cut out of the same tree be to this as 5 is to 4·714, yet the strength is less in the square beam as 101 to 110.

The following diagram (*fig.* 28.) shows the mode of describing the form of this beam on the end of a round tree,

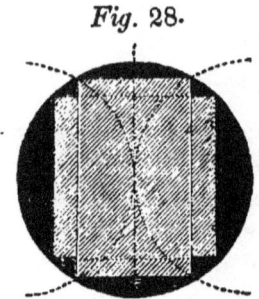

Fig. 28.

to be sawn into a crane jib. It also shows the difference between it and the square beam. If the beam be not loaded in the middle, but have the weight laid upon it nearer to one support than the other, the stress will be as the product of the two distances from the supported ends. Thus, if a beam supported at both ends be 12 ft. long between the supports, and be loaded at 4 ft. from one end, the stress upon it, compared with a similar beam loaded on the middle, will be as $4 \times 8 = 32$ is to $6 \times 6 = 36$, the stress decreasing in this proportion as the weight is brought nearer the end.

If the load be distributed equally over the whole length of the beam, the stress is equal to half that of a weight laid upon the middle of it.

If it were possible to fix the beam fast at both ends, instead of supporting it, as for instance, by building both ends into massive walls, the strength would be increased in the proportion of 3 to 2, and the beam, if loaded until it broke, would break in three places instead of one. To fix a beam in this way is practically impossible; but it should be remembered that a long joist passing over several supports, and secured to them all, benefits by this law, and is stronger than if it were cut into short lengths.

If a beam be supported at both ends in a slanting direction, at an angle, its strength, as compared with that of a horizontal beam, is as the length of the beam is to the cosine of its elevation, so that, if the length of the beam be represented by the number 5, the elevation of its end by 4, and the cosine of elevation by 3, the strength of the slanting beam compared with that of a horizontal one is as 5 to 3, and it continues to increase as it is elevated from the horizontal position, where it is least, to the vertical position where it is greatest. When it has reached the perpendicular, there is no transverse stress, for the beam has become a pillar and acts upon the thrust.

When horizontal rectangular beams are fixed at one end, and the load is suspended from the other, the strength is as the square of the depth multiplied by the breadth and divided by four times the length.

Suppose one end of the beam to be built into a wall, and a weight placed upon the other end, then suppose the beam to be lengthened and carried through the wall to an equal distance on the other side, and to be balanced by a weight at the other end, then the beam will be twice the length, loaded by twice the weight, tending to break it in the middle; so that it comes to the same result whether we divide by four times the length, or multiply by four times the weight, for a beam projecting from a wall, or otherwise secured at one end and loaded at the other, will carry only one-fourth of the weight that it will bear on the middle when it is supported at the ends. If the end of the beam be elevated above the horizontal position, its strengh increases as it approaches the perpendicular in the proportion before mentioned. The lateral or transverse strength of square beams of equal lengths is as the cube of their depth, and that of cylindrical beams as the cube of their diameter. But the strength of a round beam, compared with a square one of equal section, is as 845 to 1000; and the strength of a hollow cylinder or tube as compared with a solid one of equal section, that is to say, containing the same quantity of material, is as the difference between the fourth powers of the exterior and interior diameters of the tubular beam, divided by the exterior diameter, is to the cube of the diameter of a solid cylinder, the quantity of matter in each being the same.

All these rules are based on the assumption that the lateral or transverse fracture of a beam begins at the side or edge opposite the stress, and continues until the beam is broken through, no regard being paid to compression on the side subject to stress; and this will be found practically correct. In order to apply the rule, it is requisite to know the relative strength of the material employed, which must be ascertained by experiment. But, as most of the materials in general use have had their strength already tried, it will only be requisite to prove such as are new or unknown, and, for the strength of others, to refer to the tables in which former experiments have been reduced and classed.

The principles which govern the strength and stress of beams, as stated in the foregoing rules, are equally applicable to all materials; but in the employment of cast-iron, which may be moulded and formed into any shape or section, it would be a waste of material to make the beams rectangular or of uniform depth for the whole of their length.

The metal should be disposed according to the stress the beam has to sustain at different parts, so that, where there is most strain, there should be most material to resist it. The weight of a beam or of a crane must be taken into account as part of the load it has to carry, and metal misapplied is money thrown away.

Mr. George Rennie, Mr. T. Cubitt, Mr. Hodgkinson, Professor Barlow, Mr. Fairbairn, and others, have made many valuable experiments on the strength of cast-iron beams, and Mr. Hodgkinson has demonstrated that the best form of section for a beam carrying a load at the middle of its length between two supports is that of the annexed figure, in which two-thirds of the sectional area are given to the lower flanch. The weight that will just break such a beam is thus calculated:—

$$W = \frac{2}{3} \frac{B D^3 - (B - b) d^3}{D L}$$

Fig 29.

W = the breaking weight in tons.
L = the length in feet, 20.
B, b, D, d = the breadths and depths shown in the figure, all in inches.

No account is taken of the top flanch in this calculation—in fact, it adds so little to the strength of the beam that it is practically of no value in that respect; but it is necessary, in casting the beam, to make it with such a flanch, in order to prevent its distortion in cooling, and also to insure the soundness of the rib at top, that it may serve as a fulcrum to the lower portions of the beam which act by tension.

Thus, if the beam have the following dimensions, namely, the breadth of the lower flanch 12 in., and its thickness 2 in.; the depth of the beam 10 in., and the thickness of the web or middle part 1 in.; the width of the top flanch 5 in., and its thickness also 1 in.; then the whole sectional area of the beam will be 36 in. at the centre, and the lower flanch will contain 24 in., or two-thirds of the section. Suppose the beam to be supported at both ends, and to be 20 ft. long between the supports, then—

```
The cube of the depth or D³ is   . .  =     1000
The breadth  . . . . B . . . =               12
The reduced breadth B — b = 11            12000
The reduced depth d³ or 8³ = 512 × 11 =    5632
The length L, or 20 feet, × D = 10 or  200)6368
Of which take ⅔, which will be equal to   3)31·84
    the ultimate weight W that the beam     10·61
    will bear in tons.                          2
                                      Tons 21·22
```

that is to say, the load that will just break the beam. Mr. Hodgkinson also, in his evidence concerning the fall of a cotton mill at Oldham, gave the following approximate rule for beams of similar section.

Multiply the area of the section of the lower flanch in the middle of the beam by the depth of the beam there, both in inches, and divide the product by the length of the beam between its supports in feet; the quotient multiplied by 2·14 will give the breaking weight in tons. This rule applied to the same beam will give the following result:—

```
Area of the lower flanch 12 × 2 =    24 inches.
Depth of the beam in the middle      10
Divided by the length in feet . .  20)240
                                      120
Multiplied by the given number       2·14
Gives as the weight to break it    25·68 tons.
```

The author has found by many experiments on cast-iron of good quality, that rectangular bars supported at both ends, when reduced to a general standard of an inch square and a foot long, broke with 1 ton on the middle of the bar.

If, therefore, the same beam be supposed to form part of a rectangular parallelogram, at its middle section, which is 12 in. broad and 10 in. deep, and from which two beams, each $5\frac{1}{2}$ in. broad and 8 in. deep, be cut out, disregarding in this case also the top flanch, the following will be the result. (See *fig.* 29.)

$$10^2 \times 12 \ldots \ldots = 1200$$
$$8^2 \times 5\tfrac{1}{2} \times 2 \ldots = 704$$
Divide by length in feet $\quad 20)496$
Weight that just breaks it $\quad \overline{24\cdot 8}$ tons.

In no case should a cast-iron beam be loaded with more than one-fifth of the weight that will break it, even where the load is quiescent, as in the floors of warehouses. It has been the practice of some persons in building cotton mills to load the beams with one-third of the breaking weight, but in time they often give way, and cause frightful accidents.

The fall of a cotton mill at Oldham, in the county of Lancaster, from the failure of the cast-iron beams and pillars, and some other casualties of a similar nature, caused an inquiry to be made under a Royal Commission by Sir Henry de la Beche and Thomas Cubitt, Esq., which elicited some very important evidence, and induced some valuable experiments, to which the author has made frequent reference.

Cast-iron cranes should not be loaded with more than one-tenth of the weight that will break them, and when those cranes are intended to lower heavy bodies by means of brakes, their load should not be more than one-twentieth of the breaking weight. When a weight is rapidly lowered, and checked in its descent by the brake, it acts like a falling body upon the chain at the end of the jib. A civil engineer of acknowledged talent designed several cast-iron cranes for shipping blocks of stone; they were found quite strong enough to hoist the intended load, but failed when the blocks of stone were suddenly stopped in their descent into the hold of the ship.

Besides, it often happens, when heavy objects lie beyond the sweep of the crane, that the chain is carried out beyond the perpendicular, and the jib is thus virtually lengthened.

On beams loaded at the middle the stress decreases towards the ends, and at the points of support the beam requires, theoretically, no depth: this diminution of depth from the centre to the ends is proportionate to the square root of the stress at each point, and, consequently, forms parabolic curves, terminating when they reach the supports; practically, however, they cannot so terminate; but, when the beam is made half the depth at the ends that it is in the middle, it includes the parabolas. Or, if the beam project from a wall or a crane post, and be loaded at the end so that fracture shall begin at the upper side, the stress on the beam next to the wall will be four times the load upon the end.

If the depth of the beam next to the wall or prop be represented by 10, and the stress at the end by 100; and if the length of the beam be divided into 8 parts, then the stress upon the beam and the depths required to resist it will be in these proportions.

Fig. 30.

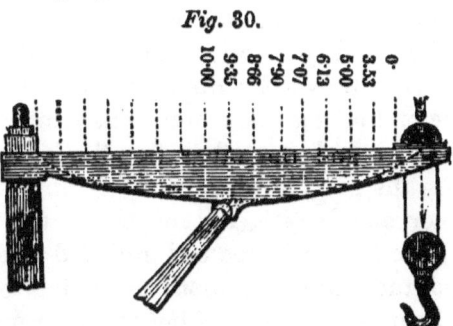

The same principle is often carried still further by casting the beams, or crane jibs, like a piece of open framework, by which the effect, if increased, may to some extent be obtained, as is shown in the construction of the 10-ton crane. (See *Fig.* 12.)

When cast-iron is subject to tension, the strength is in proportion to the areas of the transverse section of the bars, but it is seldom applied to resist tensile stress, wrought-iron being generally used for that purpose on account of its greater toughness and strength.

It is, therefore, only necessary to state that the mean of many experiments shows that the average tensile strength of cast-iron is nearly 7·2 tons to each square inch of section.

When cast-iron is subjected to a crushing action, the force required to crush a prism of a height varying from 3 to 6 times its radius is on the average $6\frac{1}{2}$ times as much as will tear it asunder, or about 48 tons upon a square inch. When the height exceeds 3 times the diameter of a solid cylindrical column, it is partially bent as well as crushed, and, when its height is less than $1\frac{1}{2}$ times its diameter (or rather 1·42), the portions crushed cannot detach themselves, because the material splits off at an angle of 55′ with the plane of the crushing surface. But, when cast-iron is used for pillars, many circumstances must be taken into account which materially affect the results, for not only are the crushing and bending actions combined, but their effects vary with their peculiar application. As, for instance, long pillars with their ends perfectly flattened, firmly fixed, bore three times as much as when their ends were rounded, and made capable of turning like a universal joint. When one end was rounded and the other flat, the strength was a mean between the two; so that in three long pillars, all of equal diameter and length, one of them with both ends made round, another with one end rounded and the other end flat, and the third with both ends flat, the strength was as 1, 2, 3, or very nearly so; but the enlargement of the flat ends with a disc beyond the diameter of the pillar, so as to give increased breadth of bearing, although necessary in many cases for practical purposes, gives no additional strength to the pillar, but stability only.

A long pillar with both ends flat, or firmly fixed, has nearly the same strength with one of the same diameter and half the length, with both ends rounded.

A solid pillar, enlarged at the middle, as 3 to 2, or upwards of the diameter of the ends, and tapering from the middle to the ends, like the frustums of two cones with bases united, has its strength increased more than the weight of the metal by about a seventh of the whole, whether the ends be round or flat.

In practice, however, it is proper that the form of parabolic spindles be given instead of the conical shape, for the following reasons, in respect of similar pillars—a shape something like the main yard of a ship. For, if long pillars be cast and turned perfectly similar, the diameter being to the length in a constant proportion, the strength of the larger pillars is found to increase in the ratio of the 1·865 power of the diameter, or nearly as the squares.

If a pillar have flat ends, but the pressure it sustains acts diagonally through it, that is to say, from the extremity of the diameter at one end to the opposite extremity of the diameter at the other, the strength is reduced in the proportion of 1 to 3, which has been proved by experiment, as in the case of pillars rounded at the ends.

These properties apply to pillars of such a length that fracture may be considered as having been produced wholly by the flexure of the column, that is to say, to cast-iron pillars with rounded ends, in which the length is more than 15 times the diameter; and to those with flat ends, whereof the length is more than 30 times the diameter; but, if the pillars be shorter, fracture takes place partly by flexure and partly by crushing, so that both these actions must be taken into the calculation.

Professor Hodgkinson found the strength of long cast-iron columns with rounded ends to increase as the 3·76 power of the diameter nearly, and those with flat ends as the 3·55 nearly, the length in each case being given. But when the length varied, and the diameter remained the same, he found that the strength was inversely as the 1·7 power of the length, or nearly so.

Taking, therefore, 3·6 as the mean between 3·76 and 3·55, and the coefficients obtained by a series of experiments, he deduced the following rules for columns fixed at the ends:—

For solid columns—

$$W = 44\cdot16 \frac{D^{3\cdot6}}{L^{1\cdot7}} = \text{the strength of a cylinder.}$$

For hollow columns—

$$W = 44\cdot34 = \frac{D^{3\cdot6} - d^{3\cdot6}}{L^{1\cdot7}} = \begin{cases} \text{the strength of a hollow} \\ \text{cylinder.} \end{cases}$$

W is the breaking weight in tons, D the external and d the internal diameter in inches, and L the length in feet. If both ends of the pillars be rounded, divide the result by 3. If one end be rounded and the other flat, take two-thirds of the result as to strength.

If pillars with flat ends be shorter than 30 times their diameter, or if the ends be rounded they be less than 15 times the diameter, they will be crushed as well as bent, and the value of W must be modified thus:—

$$W' = \frac{Wc}{W + \frac{3}{4} c}$$

c is the weight that would crush the pillar in tons, if it were so short as to be broken without flexure. To find c, Mr. Hodgkinson multiplies the area of the section of the pillar in inches by 49, because the iron he used (Low Moor, No. 3 Iron) required 49 tons to crush a prism whose base was one inch square.

* The relative strength of cast-iron pillars, as compared with other materials used in cranes, may be thus stated: representing the strength of cast-iron columns by 1000, the strength of wrought-iron was found to be 1745; cast steel,

2518; Dantzic oak, 108·8; red deal, 78·5. (*Philosophical Transactions*, 1840, *Part 2nd, page* 430.)

From what has been already stated, it is evident that all crane jibs, acting on the thrust, should have their diameter largest on the middle and tapering in a curved form, approximating to the parabolic, toward the ends, which should be about ⅔ the diameter of the middle, and also in order to obtain the greatest strength with the least material, in a cast-iron crane jib, acting on the thrust, it should be made hollow, and the ends should be fixed, so that the stress shall be taken directly through the axis of the pillar, or that the ends of the pillar shall be flat, and their planes at right angles with its axis. So that all the contrivances of ball and socket ends, and other methods of compensating for imperfection and want of truth in the construction and adjustment of pillars and struts, should be abandoned, as worse than useless.

The following tables, delivered in evidence by Mr. Hodgkinson and Mr. Fairbairn, show the results of many experiments made by them on the tensile, crushing, and transverse strength of cold and hot-blast iron.

FORCE IN POUNDS REQUIRED TO TEAR ASUNDER A BAR OF CAST IRON, ONE INCH SQUARE.

Description of Iron.	Cold Blast.	Hot Blast.	Ratio of Strength, that of Cold Blast Iron being 1000.
Buffery, No. 1.	17466 (1)*	13434 (1)	1000 : 769
Carron, No. 2.	16683 (2)	13505 (2)	,, : 809
Coed Talon, No. 2.	18855 (2)	16676 (2)	,, : 884
Carron, No. 3.	14200 (2)	17755 (2)	,, : 1250
Devon, No. 3.	,.....	21907 (1)	,, : ,,

* The numbers between parentheses show the number of experiments.

FORCE IN POUNDS REQUIRED TO CRUSH A PRISM; THE BASE ONE INCH SQUARE; THE HEIGHT, 1½ INCH.

Description of Iron.	Cold Blast.	Hot Blast.	Ratio of Strength, that of Cold Blast Iron being 1000.
Buffery, No. 1. . . .	93366 (4)	86397 (4)	1000 : 925
Carron, No. 2. . . .	106375 (3)	108540 (2)	,, : 1020
Coed Talon, No. 2. . .	81770 (4)	82734 (4)	,, : 1012
Carron, No. 3. . . .	115442 (4)	133440 (3)	,, : 1156
Devon, No. 3. . . .	......	145435 (4)	,, : ,,

TRANSVERSE STRENGTH OF BARS, ONE INCH SQUARE, LAID ON SUPPORTS 4½ FEET ASUNDER, AND BROKEN BY A WEIGHT IN THE MIDDLE.

Description of Iron.	Cold Blast.	Hot Blast.	Ratio of Strength, that of Cold Blast Iron being 1000.
Buffery, No. 1. . . .	463 (3)	436 (3)	1000 : 942
Carron, No. 2. . . .	476 (3)	463 (3)	,, : 973
Coed Talon, No. 2. . .	408·7 (3)	409·2 (2)	,, : 1001
Do. do. No. 3. . .	538 (2)	496 (2)	,, : 922
Carron, No. 3. . . .	444 (3)	520 (3)	,, : 1170
Devon, No. 3. . . .	448 (2)	537 (2)	,, : 1190
Muirkirk, No. 1. . . .	444 (2)	418 (2)	,, : 942
Elsicar cold and Milton hot blast, No. 1. . .	430 (2)	352 (2)	,, : 819

Wrought iron is seldom used on the thrust in constructing cranes; but its strength, compared with that of cast-iron, applied as a pillar, is nearly 75 per cent. greater. It has, however, been found, by experiments carefully made, that it is permanently compressed with about 11 tons on the square inch of transverse section; with loads below that weight, the bar or pillar regained its original length. It is, almost invariably, in large cranes, upon the stretch, and mostly in the manner shown in the 5-ton crane. (See Fig. 11.) Its tensile strength is directly proportionate to the area of its transverse section.

A series of experiments were tried by the late Mr. John Kingston, in the dockyard at Woolwich, which are given in all their details in the Transactions of the Society of Arts, Volume 51, 1837.

				Inches.
A bar 1¼ inch with	5	tons stretched	·011 in 100
,, ,,	10	,, ,,	·054 ,,
,, ,,	15	,, ,,	·110 ,,
,, ,,	26·5	,, ,,	1·35 broke.
A bar 1 inch diam.	5	tons stretched	·025 in 100
,, ,,	10	,, ,,	·060 ,,
,, ,,	14	,, ,,	·093 ,,
,, ,,	23	,, ,,	2·502 broke.
A bar 1 inch square	11	,, ,,	·875 {elasticity destroyed.
,, ,,	12	,, ,,	1·03 ,,
A bar 1 inch diam.	9	,, ,,	1·03 ,,
A bar 2 inches sq.	40	,, ,,	1·05 ,,
,, ,,	40	,, ,,	·90 ,,

These and various other experiments show the breaking strength of bar iron to vary from 23 to 28 tons per square inch of section, and that the average weight which breaks the bars, when they are of a fair quality, is about 25 tons for each square inch of section; that it is permanently stretched and its elasticity destroyed by about two-fifths of the strain that breaks it, or about 10 tons, the strain varying from 8·25 to 12 tons.

It has also been ascertained that bar iron stretches, within the limits of its elasticity, about ·000096, or one ten-thousandth part of an inch, by each ton of strain, and it is permanently stretched when the extension reaches one-thousandth part of its length. Consequently, it should never be subjected, for any practical purpose, to a stress of more than 5 tons on each square inch of section with a quiescent load or steady pull.

In cranes where tension rods are used the stress should

not exceed 3 tons per square inch, and if brakes be used for lowering the load, not more than two tons per square inch of section; otherwise considerable reaction may take place when the descending load is checked suddenly in lowering.

The lateral or transverse strength of wrought iron was carefully tried by Professor Barlow, at the request of the Directors of the North Western, or London and Birmingham Railway Company. His principal experiments were made on bars about 3 feet long, supported at both ends, the distance between supports being 33 inches; and he found that bars $1\frac{1}{2}$ inches broad and 3 inches deep would bear $4\frac{1}{4}$ tons of stress at the middle of the bar, as their ultimate load; with any further load the bar was permanently bent, and its elasticity destroyed. The ultimate deflection in one case was ·148 inches, and the mean deflection for each half ton of load ·0103 inches; in another it was ·124, and the mean for each half ton ·0108 inches.

The author's practice gives the lateral strength of wrought iron, as compared with cast iron, to be about 14 to 9, and in the disposition of the metal the same rules apply to both; but wrought iron cannot be run into moulds, nor can the same forms always be given to it by forging.

Fig. 31.

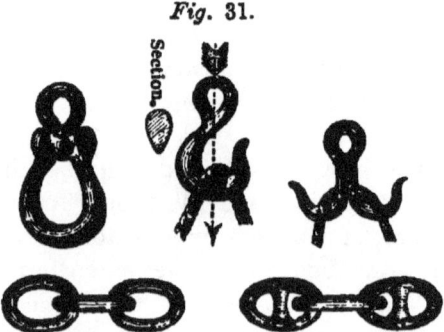

There is one part of a crane subject to severe transverse strain, that is too often neglected by the engineer and left to the smith; more lives are lost and more goods are damaged by the breaking of the crane hook than by any other part of the machine. The hook at the end of a crane chain may frequently be seen formed by simply bending a bar of round iron into a double crook or "ram's head," a shape not at all calculated to resist the stress to which it is liable. The proper form and disposition of material are shown in Fig. 31, contrasted with the older shape of the inverted "Aries;" but, when very heavy weights must be lifted, it is best to substitute a shackle for a hook. The same remarks apply to the fastening of the chain at the end of the crane jib, and also to the crane barrel.

The Deflection of Materials

To be taken into Account in constructing Cranes, or in calculating their Strength.—Timber acting on the Thrust.

In what has been stated respecting the strength of materials, their deflection has not been taken into account; and, on the construction of a crane, the form given to cast iron to afford the requisite strength is also generally sufficient to give the required stiffness.

But, in a course of experiments on cast-iron beams, the author found that a beam of 4 ft. long, which broke with $12\frac{3}{4}$ tons, deflected one-eighteenth of an inch with a load of $12\frac{1}{4}$ tons; whereas a similar cast-iron beam, 9 ft. long, which broke with a load of 13 tons, deflected half an inch with $12\frac{1}{2}$ tons.

From a series of experiments, tried on beams or battens of deal, he was led to the conclusion that, although the strength of a rectangular beam be as the square of the depth, multiplied by the width, and divided by the length multiplied into the load laid upon the middle of the beam, when supported at the ends, yet that the deflection of the beam increases directly in proportion to the load and to the cube of the length, and inversely as the cube of the depth multiplied by the width, a rule confirmed by the experiments of Professor Barlow. It is, therefore, important in all works, constructed on a large scale, that the deflection of the material used shall enter into the calculation, but more especially in those cases in which timber may be employed in the form of rectangular beams, or some combination of them.

In order to do this, it is requisite to know what amount of deflection the material will bear, and what load will produce that deflection, so that it may be kept within proper limits.

The elaborate experiments of Professor Barlow, made by order of the Admiralty in the dockyard at Woolwich, and reduced by him into the form of tables in his excellent work on "The Strength of Timber and other Materials,"[*] afford data for calculating the dimensions of most kinds of wood likely to be used in works of strength.

These tables are the more valuable, as the experiments were numerous and exact, and the timber on which they were made was of considerable scantling, the pieces being generally 7 ft. long and 2 in. square; but, as the wood was of the best quality, well seasoned, and free from knots, sap, and other defects, allowances must be made in applying the results to practice, to compensate for such imperfections.

The beams were fixed at one end and loaded at the other; the greatest amount of deflection they could sustain without injury was taken as the elastic strength of the particular kind of wood under examination, and then the load was increased until the beam broke.

It is obvious that, when the elasticity of the beam was destroyed, or, as a carpenter would say, "when it was crippled," it became useless to pursue the inquiry further, as affording data for calculation; and it appears to be a good general rule that the greatest load to which the beams should be subjected is one-fourth of the weight that ultimately breaks them.

These experiments also prove that the deflection of a

[*] In 8vo, edition 1851.

ON THE CONSTRUCTION OF CRANES. 83

beam, fixed at one end and loaded at the other, is to that of a beam of the same length, supported at both ends, and loaded at the middle with the same weight, as 32 to 1; the stress, as has before been shown, being as 4 to 1.

The elastic strength of the timber, that is to say, the greatest load it will bear without having its elasticity destroyed, he represents by the letter E. The ultimate strength of the timber, that is to say, the stress that breaks it, he represents by the letter s, and has arrived at the value of these letters in the following manner:—

A beam of teak timber, 2 in. square, fixed at one end and projecting horizontally 7 ft. from its support, was loaded on the other end, and by careful and repeated trials, such wood, which in this experiment had the specific gravity of 745, had its strength and deflection ascertained, the load being gradually increased until the beam broke. The mean results were as under:—

Greatest weight sustained while the elasticity
    remained perfect in lbs. 300
Deflection at that time, in inches . . . 1·151
Breaking weight, in lbs. 938
Ultimate deflection before fracture, inches . 4·320

To find the value of E, cube the length in inches, namely, $84^3 = 592,704$, and multiply this by the load in pounds $= 300$, which gives a product of 177,811,200.

Divide this product by the breadth, 2 inches, multiplied into the cube of the depth, or $2^3 = 8$, multiplied into the deflection 1·151, multiplied by 32, or a total divisor, $= 589,312$, and the quotient will be 301,800, very nearly; this is the value of E, which, for the sake of simplifying operations, is not carried out exact beyond the fourth figure.

The value of s is thus found:—Multiply the length in inches, 84, by the breaking weight in lbs., 938, and their product is 78,792.

Divide the product by that of the breadth, 2 in., multiplied into the square of the depth, = 4 in., multiplied by 4, or a total divisor of 32, and the quotient will be 2462, which is the value of s, and represents the ultimate strength of the timber.

All the dimensions are here taken in inches; but, if the length be taken in feet, the number 301,800, which represents the value of E, must be divided by the cube of 12 in., or 1728; this reduces it to 174, and represents the load in pounds that a piece of teak, 1 in. square, and 1 ft. long, will bear when supported at both ends, and weighted on the middle, without destroying its elasticity.

In like manner, if the number 2462, which represents the value of s, be divided by 12 in., it is reduced to 205, and shows the weight in pounds that will just break the same piece of wood.

Practical men have great objections, and very properly so, to the use of tabular numbers, when they are arbitrarily given; but the experiments, in this instance, being given with them, enable persons to judge of their applicability to practice, and to make allowances for accidental defects in the timber they have to employ.

The following table shows the elastic strength (E) and the ultimate strength (S) of those kinds of timber in general use, which are calculated by this formula:—

$$\frac{l^3 \times w}{b \times d^3 \times 32 \times D} = E; \text{ and } \frac{l \times w}{b \times d^2 \times 4} = S;$$

l being the length, b the breadth, d the depth, w the load or weight carried, and D the deflection; all the dimensions being taken in inches.

If a beam of Riga fir, 4 in. broad and 6 in. deep, be fixed at one end, and project 5 ft. over the point of support, and be loaded at the end, if it be required to determine what is the greatest load it can bear without injury, deflection not being considered, find the value of s for Riga fir, which is 1108, multiply this by the breadth and by the square of the depth, and divide the product by 60 in., the

Names of the Woods. The Pieces tried are all 84 Inches long and 2 Inches square, except those otherwise marked.	Specific Gravity.	Greatest Weight and Deflection while the Elasticity remained perfect.		Breaking Weight in Pounds.	Ultimate Deflection in Inches.	Value of E from Formula $\dfrac{l^3 \times W}{b \times d^3 \times 32 \times D} = E.$	Value of S: $\dfrac{l \times W}{b \times d^2 \times 4} = S.$
		Weight in Pounds.*	Deflection in Inches.				
English Oak	745	300	1·151	938	4·32	301800	2462
Do. do. inferior wood	934	200	1·280	637	8·10	181400	1672
Canadian Oak	969	150	1·590	450	5·90	109200	1181
Dantzic Oak	872	225	1·080	673	6·00	268600	1766
Ash	756	200	1·590	560	4·86	148900	1457
Beech	760	225	1·266	772	8·92	205600	2026
Elm	696	150	1·026	593	5·73	169200	1556
Pitch Pine	553	125	1·685	386	6·93	87480	1013
Red Pine	660	150	1·134	622	6·00	153200	1632
New England Fir	657	150	·755	511	5·83	230000	1341
Riga Fir	553	150	·931	420	4·66	273900	1102
Do. do. (6 feet long)	753	125	·870	422	6·00	166100	1108
Mar Forest Fir	738	150	·883	467	6·00	123800	1051
Do. do. (6 feet long)	696	125	1·442	436	6·00	80670	1144
Larch Fir (6 feet long)	703	150	1·006	561	6·42	108700	1262
Do. do.	560	150	·831	510	5·00	131600	1149
Norway Spar	531	125	1·885	325▼	8·58	77040	853
	577	200	·800	655	4·00	182200	1474

length of the beam. The result gives the greatest or breaking load, one-fourth of which is as much as it will safely carry. Thus:—

$$\frac{1108 \times 6 \times 6 \times 4}{60} = 2659 \cdot 2 \text{ lbs}$$

the breaking load, and

$$\frac{2659 \cdot 2}{4} = 664 \cdot 8 \text{ lbs.}$$

the greatest load the beam will safely carry.

If a beam of Riga fir, 1 ft. square and 20 ft. long, be supported at both ends, and loaded in the middle, the weight it will sustain is found by multiplying 1108 by 4, and proceeding as before. Thus:—

$$\frac{1108 \times 4 \times 12 \times 144}{240} = 32010 \text{ lbs.}$$

which is the breaking load. One-fourth of this is the greatest weight the beam will bear without injury, namely 8002 lbs.

If the beams be inclined, as they often are in cranes, the leverage or projection taken on the horizontal line, or the distance between the supports measured horizontally, must be considered as the length of the beam. If the load be spread uniformly over the whole length of the beam, it must be reckoned at half its weight; and in no case should the load be greater than one-fourth of the breaking weight, besides allowing amply for imperfections, and for deterioration by time and weather in exposed situations.

When it is required to determine the dimensions of a beam capable of supporting a given weight with a given degree of deflection when fixed at one end,

To lessen the number of figures, reduce the value of E by dividing it by 1728, the cube of 12 in., which brings the length of the beam into feet. Divide the load in pounds by this reduced value of E multiplied by the breadth and deflection, both in inches; then the cube root of the quotient, multiplied by the length in feet, will be the depth required in inches. Thus, a beam of Riga fir is intended to bear a load of 665 lbs. at one end, when it is fixed at the other. The length of the beam is 5 ft., the breadth is 4 in., and the deflection is not to exceed ¼ of an inch.

The value of E, 166,100, divided by 1728, is 96 nearly; hence

$$\frac{665 \text{ lbs.}}{96 \times 4 \times \cdot 25} = 6\cdot 88,$$

the cube root of which is 1·902, and 5 × 1·902 = 9·51 in., the depth of the beam required.

It has been shown that a beam of 6 in. in depth would carry the load; but where a certain stiffness is requisite, as well as strength, it becomes necessary to adopt this mode of calculation.

If the beam be cylindrical, as a mast or spar, the deflection will be to that of a square beam as 1·7 to 1, other circumstances being the same.

When it is required to find the size of a beam supported at both ends, capable of supporting a given weight, with a given degree of deflection, multiply the load to be carried in pounds by the cube of the length in feet. Divide the product by 32 times the reduced value of E, multiplied into the given deflection in inches, and the quotient is the breadth, multiplied by the cube of the depth in inches.

If it be intended to make the beam square, then, the breadth and depth being equal, the fourth root of the quotient is the side of the square.

If the proposed beam be round, first multiply the quotient by 1·7, and then extract the fourth root as before, which will be the diameter of the cylindrical beam sought.

A cylindrical beam of Riga fir, 20 ft. long, supported at both ends, is required to carry in the middle a load of 7 tons, and the deflection must not exceed the ·2, say two-tenths of an inch.

ON THE CONSTRUCTION OF CRANES.

In the former case, the tabular value of E was at once divided by the cube of 12 inches = 1728; in the present instance, for the sake of being more exact in dealing with large numbers, and very small deflections, multiply the value of E 166,100 by 32, and then divide by 1728, which gives 3075, and reducing 7 tons to pounds gives 15,680.

$$\text{Then } \frac{1{\cdot}7 \times 15680 \times 20^3}{3075 \times {\cdot}2} = 346744$$

The fourth root of this number, 24·26 inches, is the diameter of the beam required.

```
Thus 5 | 346744 (☞         588·85(24·26
     5 |  25                 4
     ───────                 ──────
   108 |    967             44 | 188
     8 |    864              4 | 176
    ──────                   ──────
  1168 |  10344             482 | 1285
     8 |   9344               2 |  964
  ──────                     ──────
 11768 | 100000             4846 | 32100
     8 |  94144                  | 29076
 ──────                          ──────
117765 | 585600
       | 588825
```

Many cranes are now so constructed, that a pillar or strut of timber acting upon the thrust forms the jib, and two tension bars of iron, extending from the point of the jib to the top of the crane-post, or its framework, act as stays. It is, therefore, important to know what stress a pillar of wood will support in the direction of its axis without sensible deflection. Unfortunately, there are few direct experiments to show the precise strength of timber so applied, and many, who have written upon this subject without data, are evidently wrong in their results. Experiments which show the force required to crush short pieces, or cubes of wood, are manifestly useless for calculating the strength of long pillars made of a flexible material like timber The practical experience of the author has led him to the following conclusions:—First, that the pillar, or strut, must be fixed at the ends, and applied directly to the thrust, so that the stress may be exactly in the direction of the axis, and, consequently, that circular ends, rule joints, ball and socket terminators, and all contrivances of that kind, are detrimental, as, indeed, Mr. Hodgkinson's experiments prove. Next, that there shall be no sensible flexure in the pillar, and no tendency to bend more on one side than another; therefore, the strongest form is that of a square beam; and, lastly, that the value of E, or the power of resistance to compression in wooden pillars, liable to give way by flexure, is not greater when applied to practical purposes than that of Mr. Barlow's tables multiplied by 10. Assuming this to be correct in practice, the strength of wooden pillars will be directly as the fourth power of their diameters or thickness, and inversely as the squares of their length. These proportions approximate very nearly to those given by Mr. Hodgkinson for iron pillars. Hence may be deduced the following rule:—

Square the length of the pillar in feet, and multiply by the load in pounds. Divide the product by 10 times the value of E, in Mr. Barlow's tables, reduced for feet, (by dividing by 1728,) and the quotient will be the fourth power of the thickness or side (in inches) of a square beam, which will support that thrust.

If the pillar be round, multiply the quotient by 1·7, and then find the root. For instance, a Riga spar is required to bear a direct thrust of 5 tons, and its length is 20 ft.; what diameter should it be?

5 tons are equal to 11200 lbs.
20 feet squared, equal to 400.
E = 166100 divided by 1728 = 96 × 10 = 960.

Then $\dfrac{11200 \times 400}{960} = \sqrt[4]{4666} = 8·2$ inches square;

for a round pillar, $4666 \times 1·7 = \sqrt[4]{7932} = 9·4$ inches diam

Thus the jib of a crane, to sustain a thrust of 5 tons in

the direction of its length, if made of a Riga spar, should be 9·4 in., or say 9½ in. in diameter at the middle of its length. If the spar be short in proportion to its diameter, that is to say, if the length do not exceed 15 diameters, it may be diminished one-fourth at the ends, without appreciable loss of strength, the diminution, or taper, being the frustum of a parabolic spindle. If it be 18 diameters, the diminution at the ends may be one-fifth, and, if 21 diameters, one-sixth of the diameter. This taper is similar to the Entasis of the Greek column.

The large sheers for masting ships, and for lifting the heavier parts of marine engines, erected at Woolwich Dockyard from the design of Oliver Lang, Esq., master-shipwright, present a good example of resistance to stress, exerted in the direction of their length, that is to say, acting on the thrust. (See Fig. 32).

The centre mast is built somewhat like the mainmast of a line-of-battle ship; much skill is shown in the arrangement of the pieces, and the mode of securing them by means of plates and hoops of iron, the whole being so firmly put together, that it seems to realise the adage of a bundle of sticks that could not be broken when so combined.

The size of the spars required for the sheers and sprits, both as to diameter and length, but especially the latter, was so great, that it would, perhaps, be impossible to procure timber of natural growth fit for the purpose. The ends of the spars composing these "sticks" are combined in a peculiar manner, so that a straight and strong piece is artificially formed of parts which are probably strongest at their junction. A thick octagonal plate of iron is let into the end of each spar where the parts join, and four strong dowels of iron, 2 ft. long and 4 in. in diameter, pass through these plates into the timber. Between these two plates is a third and thicker plate, somewhat larger than the ends of the spar, through which also the dowels pass; it is notched or "scored" on the edge to receive long iron splints. sur-

ON THE CONSTRUCTION OF CRANES. 91

Fig. 32. MASTING SHEERS

At Her Majesty's Dock Yard, Woolwich;

Constructed by the late Oliver Lang, Esq., Master Shipwright.

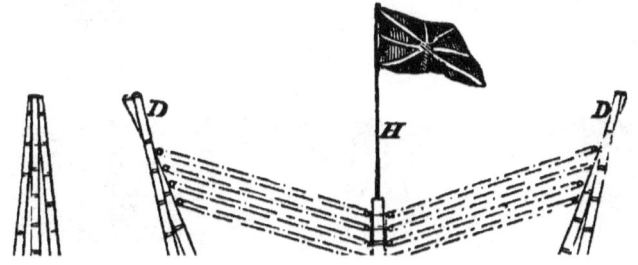

rounding the spar, and firmly hooped round and bolted through the wood, thus completely securing the junction

Fig. 33.

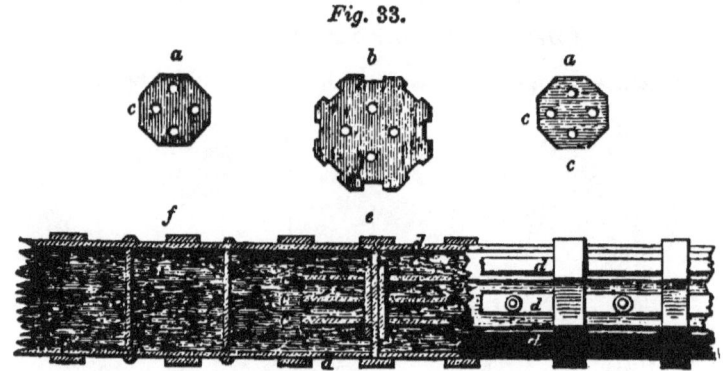

a a, are octagonal plates of wrought iron, 1½ inches thick, let into the abutting end of each spar.

b, the centre plate, 1⅞ inches thick, through which, as well as through the other octagonal plates (*a a*), the iron dowels pass, to strengthen the butts or junction of the spars.

c c, the wrought-iron dowels, four in number, each 2 feet long and 4 inches in diameter.

d d, iron straps or splints for connecting the spars together, let in flush, and bolted through the spars and through each other.

e, a broad thick hoop drawn over the centre plate to confine the straps in the notches on the edge of the plate.

f f, additional hoops to bind together the wrought iron straps and the timber.

Of the Stress or Force acting upon each Part of a Crane.

In order rightly to apply those materials, the strength of which has been considered, it is requisite, in the construction of a crane, to lift or lower a heavy load, to determine how that load will act upon the different parts composing it; and, therefore, briefly to show how the load or force may be resolved in a practical way.

The usual form of a crane for general purposes is that of a right-angled triangle, the three sides being the stalk or upright, the jib or arm, and the stay, which is sometimes also called the spur or strut, and is the hypothenuse of the triangle. (See Fig. 34.)

When the jib and the stalk are equal in length, and the stay is the diagonal of a square, the form is theoretically strongest, because the whole weight acts upon the stay, tending to compress it in the direction of its length, the stress upon it, as compared with the weight, being as the diagonal to the side of the square, that is to say, as 1 to the square root of 2, or as 1 to 1·4142, nearly; consequently, if the weight upon the crane be 10 tons, the thrust upon the stay will be 14 tons 2 cwts. 3 qrs. 7 lbs., very nearly, or as B C to C D. The weight W is sustained by the rope or chain, and the tension is equal upon both parts of it, that is to say, on the two sides of the square, A B and C D; consequently the horizontal arm or jib has no stress upon it, and serves merely to retain and steady the diagonal stay B C.

If the foot of the diagonal stay be lowered to G, the thrust or compression, as compared with the weight, will be as B G to B W; and if the chain from A to B be then removed, and the weight be *suspended* from the point B, the tension on the jib will be as A B to B W.

But if the foot of the stay be raised so as to form the diagonal B F, the thrust upon it will be as the line B F is to B D; and the tension on the jib will be represented by the line E B.

By dividing the line representing the weight into equal parts, to represent hundredweights or pounds, and using it as a scale, the stress upon any other part may be measured upon the described parallelogram.

In Fig. 35, also, the angles A B E and E B C being equal, the chain or rope is represented by A B C, and the weight by W; the stress upon the stay, as compared with the weight, is as B D to A B or B C; therefore a slight rod of iron, extending from A to B, is all that is necessary to steady and retain the spur or strut, so long as the chain or rope holds good.

In practice, however, it is not prudent to consider the chain as supporting the stay; but it is proper to disregard the chain or rope as forming part of the system, and the crane should be calculated to support the load independent of it. It is also proper that the angles on each side of the diagonal stay, in this case, should not be equal. If the side A B be formed of tension-rods of wrought iron, lower the point A so as to lengthen that side and decrease the angle A B E; but if it be of timber, raise the point A and increase the angle: a little difference is sufficient to compensate for the friction of the pulleys, and, by slightly throwing the preponderance of load either in tension or in thrust upon the side A B, the crane will work with less tremor than it might do if it were in a state of doubtful equilibrium when hoisting or lowering the weight suspended by the chain.

Fig. 36 shows the parts composing the crane, arranged in the form of an equilateral triangle, in which the weight B D, the tension B C, and the thrust A B, are all equal to each other, the weight W being suspended from the point B.

Fig. 37 shows a form of crane very generally used; the angles are the same as in Fig. 35, and the weight suspended from it, being attached to the point B, is represented by the line B D. The tension, which is equal to the weight, is shown by the length of the line B C, and the thrust by the length of the line A B, measured by a scale of equal parts, into which the line B D, representing the weight, is supposed to be divided.

But if B e be the direction of the jib, then B g will show the tension, and B f the thrust, both of them being now greater than before; the line B D, representing the weight, being the same in both cases.

Fig. 38 shows the preceding figure, or framework, reversed, and the weight suspended from it as before; the

DIAGRAMS 95
Showing the stress on the different parts of a Crane.

Fig 34

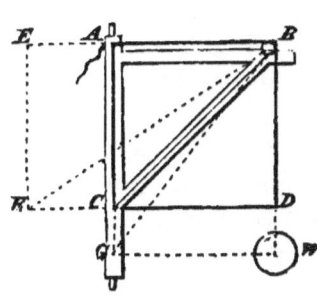

Fig 36

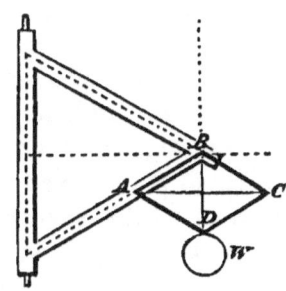

Fig 35

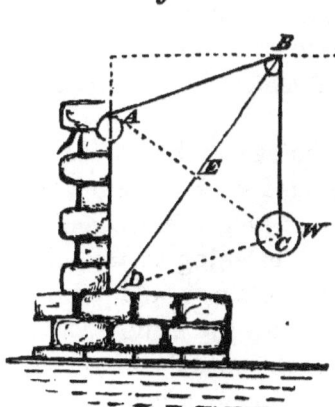

Fig 37

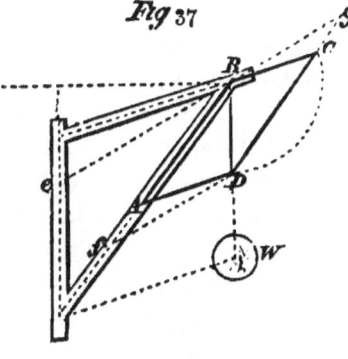

Fig 38

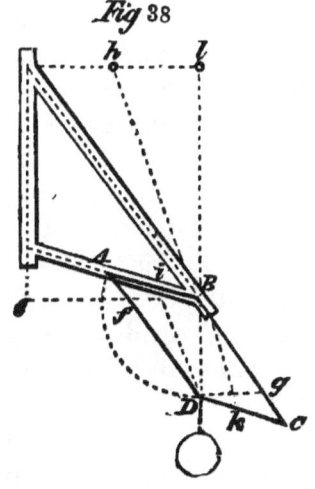

weight and the thrust are now equal, and the *tension*, in this instance, is equal to the *thrust* in the last case.　B D shows the weight, B c the tension, and A B the thrust.

But, if the thrust be taken in the direction of the line B *e*, then are both the tension and the thrust diminished, as shown by the parallelogram *f* B, D *g*.

If the tension be taken in the direction of B *h*, the diminished tension and thrust, as compared with the weight, are shown by the two short sides of the parallelogram *i* B and D *k*, the weight, as before, being represented by the line B D. Again, if the tension be taken in the direction B *l*, it forms a straight line to the weight, and no other force is exerted. Thus, by dividing the line that represents the weight into a scale of equal parts, the forces may be measured by it, the stress on each part of the crane may be determined, and the strength of materials, proportioned to the stress, may be calculated.

If the crane shown in Fig. 34, instead of being secured by pivots at top and bottom, were fixed in the ground, in a well, or set in masonry up to the level of the line c D, without support from any pivot or fastening at top, then the weight w, acting with the leverage c D, would tend to break the crane post at the point c, with the same effect as if twice that weight were laid upon the middle of a beam equal to twice the length of c D, the point c being at the middle of the beam which is supposed to be supported at both ends; the size of the beam, in all other respects, being the same as that of the crane post, and the depth being that of the line of fracture.

Or the force exerted to break the crane post, and the resistance to fracture, would be represented by the weight (w), multiplied by 4 times the length of the lever, or line c D, divided by the depth of the post squared, and multiplied by the breadth of it, and by the number equivalent to the timber's strength (s), making such allowances as are indicated by the character and condition of the wood, and the circumstances incident to its use.

On Chains and Ropes.

Their Application to Cranes and Hoisting Machinery, and relative Strength.
—Chains.—Hemp Ropes.—Wire Ropes.

Having determined the greatest weight that a crane shall carry, and proportioned the frame of it accordingly, it is necessary to decide on the size of the chain or rope which will safely sustain that weight, allowing also for wear and deterioration.

Chains are now so well and accurately made, and their soundness and quality so readily ascertained by means of the "Testing Machine," that they are much used for crane-work instead of ropes, especially out of doors; but, as there are many circumstances in which it may be requisite to use ropes, it will be proper to show the strength of both.

Chains for cranework are commonly made with short links of an oval form, the links being no longer than is necessary to permit the smith to use his tools for soundly welding the iron.

Chain cables for ships are made with the links somewhat longer, and a stud or stay is inserted in the link across its least diameter, to support the sides and prevent their collapse under heavy strain, thus giving great additional strength to the chains.

The barrels of cranes should have a groove cut spirally or screw fashion upon the cylinder, just so wide as will receive the edge of the link, allowing but little liberty or play, and so deep that the flat of every alternate link may bear evenly upon the cylindrical surface of the barrel.

The practice of the author induces him to recommend that chains above an inch in diameter should not be employed for cranework, preferring chains of smaller size, say $\frac{5}{8}$, $\frac{3}{4}$, or $\frac{7}{8}$ of an inch in diameter of the iron, and reducing the strain upon the chain by increasing the number of the parts that bear it, by means of blocks and pulleys.

Within certain limits the strength of chains may be estimated in proportion to the squares of the iron's diameter of which they are made, and the same may be said of ropes measured by the squares of their circumference, which is the general rule; but it is obvious that, when they are made very large or very small, neither the quality nor the workmanship can be so well relied upon. Iron of large diameter is less fibrous and more crystalline in the fracture, and the welding is more perfect in links of intermediate size. In ropes of large size, notwithstanding the great improvements in forming the strands and laying them together by machinery, the outside yarns of large ropes are, of necessity, subject to greater wear, and also to greater stress in work, by bending or coiling a piece of rigid cordage. Therefore, although the square proportion may be proper as a rule for proof, it is prudent to make allowance for such circumstances in the employment both of chain and rope, and, in all cases, to make the pulleys of as large diameter as can conveniently be done. The strength of the short-linked chain without studs, which is generally used for cranes, as compared with the studded chain used for cables, is nearly as 7 to 9; and it is, therefore, not prudent to subject both kinds to the same proof, when the soundness of the chains is tested. Taking chains made of iron 1 in. in diameter as the standard, crane-chain, with short links, may be proved to 14 tons, and cable-chain, with studded links, to 18 tons, which is the Navy proof for iron cables. Both kinds may safely be worked to half the strain to which they have been proved, but not to more.

Chains, when tested, are generally tried at a mean temperature; the strain is steady, and the vibration caused by the blows of a workman's hammer, given in order to detect imperfect welding of the links, is by no means equivalent to the jerks it will have to sustain when exposed, perhaps, to a temperature below the freezing point, affecting the strength of iron as it does that of all crystallised bodies.

The ultimate, or breaking strain of chain-cables, as com-

ON THE CONSTRUCTION OF CRANES.

pared with the Navy proof, is about 1·9 to 1; and, in proving chains about an inch in diameter, it is found that cables stretch about 3 ft., and short-linked chain about 4 ft., in a length of 15 fathoms. The strength of a chain cable, as compared with that of iron bolts equal in section to the chain, is about 7 to 9. So much does the stud add to the strength of the link, by preventing its collapse.

The following table is a comparison of short-linked crane-chains with ropes; it shows the size and weight of each, and the proof of the chain in tons. The ropes of the sizes given are considered to be of equal strength with the chains, which, being short-linked, are made without studs. The ropes are made with the register and press block.

Size of the Chains.	Weight in lbs. per Fathom.	Proof Strain in Tons.	Size of Rope.	Weight of Rope in lbs. per Fathom.
$\frac{5}{16}$	6	$\frac{3}{4}$	$2\frac{1}{2}$	$1\frac{1}{2}$
$\frac{3}{8}$	$8\frac{1}{2}$	$1\frac{1}{2}$	$3\frac{1}{4}$	$2\frac{1}{2}$
$\frac{7}{16}$	11	$2\frac{1}{4}$	4	$3\frac{3}{4}$
$\frac{1}{2}$	14	$3\frac{1}{2}$	$4\frac{3}{4}$	5
$\frac{9}{16}$	18	$4\frac{1}{2}$	$5\frac{1}{4}$	7
$\frac{5}{8}$	24	$5\frac{1}{4}$	$6\frac{1}{4}$	$8\frac{1}{2}$
$\frac{11}{16}$	28	$6\frac{1}{2}$	7	$10\frac{1}{2}$
$\frac{3}{4}$	32	$7\frac{3}{4}$	$7\frac{1}{2}$	12
$\frac{13}{16}$	36	$9\frac{1}{4}$	$8\frac{1}{4}$	15
$\frac{7}{8}$	44	$10\frac{3}{4}$	9	$17\frac{1}{2}$
$\frac{15}{16}$	50	$12\frac{1}{2}$	$9\frac{1}{4}$	$19\frac{1}{2}$
1	56	14	10	22

The rope of the above statement is such as is now generally made by machinery at most of the large ropeworks, but was formerly known as " Patent Rope," in which every yarn is made to bear its part of the strain; but, if common hand-laid rope be used, the proof strain must be reduced one-fourth, and in actual work the load should not, at any time, exceed one-half the proof.

The old ropemakers' rule was to square the girth of the rope in inches, which, multiplied by 4, gave the ultimate or breaking strength of the rope in hundredweights, and it was a good rule for small cordage, up to 7 in. in circumference.

The square of half the circumference was considered to represent the weight of a fathom in pounds.

There is considerable difference between the weight of common rope and that made by the register, although the girth and the number of yarns be the same in both, for large ropes made by the old method are about $7\frac{1}{2}$ per cent. heavier than those made by register, and not so strong. The strength, also, depends very much on the quality of the hemp, while the durability of the rope, and, consequently, its trustworthiness in use, are greatly affected by the quality of the tar, some of which contains an acid destructive to the hemp, so that the rope may be rendered unsafe in three years after it has been made.

In the execution of important works, it is, therefore, recommended that the ropes used shall have been recently manufactured; that they shall be made of the best Petersburgh hemp with Archangel and Stockholm tar; that the yarns shall be formed into strands, with the register plate and press block, and the strands laid together by the stranding machine, so that every part of the rope may do its share of the work.

The inventions and experiments of the late Captain Huddart, and of Mr. Chapman, have tended to make the use of such machinery so general, that there is no difficulty in obtaining the improved cordage, and much talent and ingenuity, and large capitals, have been employed to carry out and perfect their improvements.

The following table shows the mean results of three hundred trials made by Captain Huddart. It shows the relative strength or cohesive power of each kind of rope, taking, as a standard of comparison, $\frac{1}{10}$ of a circular inch, equal to an area of ·078, or nearly $\frac{1}{13}$ of a square inch, which corresponds with the old ropemakers' rule of dividing the number of pounds which will break a rope by the square of its girth in inches

It shows that ropes formed by the warm register are stronger than those made up with the yarns cold; because the heated tar is more fluid, and penetrates completely between every fibre of hemp, and because the heat drives off both air and moisture, so that every fibre is brought into close contact by the twisting and compression of the strand; the tar thus fills up every interstice, and the rope becomes a firmly-agglutinated elastic substance, almost impermeable to water. But, although rope so made is both stronger and more durable, it is less pliable, and, therefore, the cold registered rope is more generally used for cranework, where the rope must be wound round barrels or passed through pulleys.

A COMPARATIVE TABLE,

Showing the Weights required to break Cordage made by the old Method, and Cordage made by the Register.

Size of Ropes.		Made by the old Method.				Made by the Register.			
Girth.	Diameter.	Of common staple Hemp.	Per $\tfrac{1}{16}$ of a circular Inch in Area.	Of the best Petersburgh Hemp.	Per $\tfrac{1}{16}$ of a circular Inch in Area.	Cold Register.	Per $\tfrac{1}{16}$ of a circular Inch in Area.	Warm Register.	Per $\tfrac{1}{16}$ of a circular Inch in Area.
in.	in.	lbs.	lbs.	lbs.	lbs.	lbs.	lbs.	lbs.	lbs.
3	0·95	5050	561	6030	670	7380	935	8640	960
3¼	1·11	6784	554	8669	707	11165	911	11760	960
4	1·27	8768	548	10454	653	13108	819	15360	960
4½	1·43	10308	504	12440	614	16325	806	19440	960
5	1·59	13250	530	15775	631	20500	820	24000	960
5½	1·75	15488	512	18604	614	24805	820	29040	960
6	1·91	18144	504	21616	600	24520	820	33120	920
6½	2·07	20533	486	23623	559	34645	820	40554	959
7	2·24	22932	468	27342	558	40188	819	47040	960
7½	2·39	24975	444	30757	546	46125	820	54000	960
8	2·54	26880	421	32000	500	52480	820	61430	960

The proof-strain for chain cables used in the Royal Navy is about 630 lbs. on each circular inch of the iron bolt of which the chain may be made; as, for instance, a chain cable made of 1-inch iron contains in one side of the link

8 × 8 = 64 circular eighths, which, being multiplied by 630 lbs., give 40,320 lbs., or 18 tons.

In the succeeding table are shown the Navy proof for chain cables; the strain which will just break the single bolt of which each chain is made, when the iron is of average good quality, and the ultimate or breaking strength of the chain. Also the size of hempen cables to which these chains are considered equivalent in strength, and for which they are substituted in practice, with their weight in pounds per fathom of 6 ft.

The table commences with chain made of $\frac{5}{8}$ths iron, which is the smallest size usually made with studs, and ends with chain made of $2\frac{1}{4}$ inches iron, the largest chain cable at present in use.

Short-linked chain, without studs, is seldom made larger than $1\frac{1}{4}$ inch, and generally not more than 1 inch, diameter of iron.

The weights of the hempen cables are taken from Mr. Edye's tables for the equipment of ships of war, and reduced into pounds, so that, by moving the decimal point one or two places, it shows at a glance the weight of 10 or 100 fathoms.

For many purposes, and particularly for raising coal in large quantities from deep pits, flat ropes are now used. They combine great strength with flexibility; and, as the rope winds upon itself, like a ribbon, when drawing up the load, it forms a roller or barrel of increasing diameter, whilst the corresponding rope at the same time descending, unwinds from itself, and lessens the diameter of the drum, so that the two ropes at all times balance each other. This is of much importance when a rope of great length and weight is suspended in a deep mine; and, were it not for such an arrangement, great mechanical power must be expended in winding up the rope alone, or compensating machinery of a cumbrous and expensive kind must be used; whereas, by making the bodies of the drums or barrels of such size at the commencement that they shall be equal in diameter when the loads meet midway in the pit, the exertion of mechanical power is rendered uniform and equable.

Such flat ropes are made by placing four round ropes side by side, and stitching them together through all, by machinery, into a flat band. Two of these ropes are twisted or laid to the right hand, and two to the left hand, so that the tendency to untwist is neutralised, and the combined band rolls up straight and evenly.

Comparative Table,
Showing the Weight and Strength of Chain and Hempen Cables, the Chains being made with Studs.

The first column shows the diameter of the iron forming the links.—The second the weight per fathom.—The third the ultimate strength of the iron bolts.—The fourth that of the chains.—The fifth the circumference in inches of the cables, for which the chains are substituted. And the sixth the weight per fathom of the cables in pounds.

Size of the Chain.	Weight per Fathom in lbs.	Breaking Strength of the Bolt.		Breaking Strength of the Chain.		Navy Proof in Tons.	Hemp Cable equal in Strength.	Weight of Hemp Cable, lbs. per Fathom.
in.	lbs.	tons.	cwt.	tons.	cwt.	tons.	in. girth.	lbs.
$\frac{3}{4}$	24	8	7	13	4	7	$6\frac{1}{2}$	8·89
$\frac{13}{16}$	28	10	2	16	0	$8\frac{1}{2}$	$7\frac{1}{4}$	11·85
$\frac{7}{8}$	32	12	1	19	5	$10\frac{1}{4}$	8	13·90
$\frac{7}{8}$	44	16	4	26	5	$13\frac{3}{4}$	$9\frac{1}{2}$	19·36
1	58	21	8	34	5	18	$10\frac{1}{2}$	22·59
$1\frac{1}{8}$	72	27	2	48	15	$22\frac{3}{4}$	12	30·75
$1\frac{1}{4}$	90	33	10	53	11	$28\frac{1}{2}$	$13\frac{1}{4}$	39·27
$1\frac{3}{8}$	110	40	10	65	0	34	15	48·17
$1\frac{1}{2}$	125	48	4	77	0	$40\frac{1}{2}$	16	54·84
$1\frac{1}{2}$	145	56	11	90	10	$47\frac{1}{2}$	17	61·88
$1\frac{3}{4}$	170	65	12	105	0	$55\frac{1}{2}$	18	68·17
$1\frac{7}{8}$	195	75	6	120	10	$63\frac{1}{4}$	20	85·97
2	230	85	14	137	0	72	22	104·04
$2\frac{1}{8}$	256	96	15	155	0	$81\frac{1}{2}$	24	124·05
$2\frac{1}{4}$	285	108	0	195	0	$91\frac{1}{2}$	26	145·26

Recently WIRE has been used for making ropes, and they have been found to answer the purpose very well for working inclined planes, and for raising coal from the pit, as, in such cases, barrels and pulleys of large diameter are applicable; but the rigidity of the wire rope will not permit it to pass frequently over pulleys of ordinary size without injury from repeatedly bending it in opposite directions. The author has seen it employed with advantage where hemp ropes would soon have been cut to pieces, and where the weight of chains would have been inconvenient.

These wire ropes were made by Messrs. R. S. Newall and Co., of Gateshead-upon-Tyne, in the county of Durham, who have drawn up the following scale of comparison between wire ropes of their manufacture, and hempen ropes. It is also to be observed, that, by coating the iron wires with melted zinc, or, as it is sometimes called, "galvanising the iron," it may for a long time be preserved from rusting; and it may be remembered that iron wire of $\frac{1}{10}$ of an inch in diameter bears an ultimate strain of 30 tons to a square inch, which is greater than the breaking strength of iron, of any other form or section, subjected to tension.

The Table

Shows the Circumference of Wire Ropes, their Weight per Fathom, their ultimate or breaking Strength in Tons, and the working Load they will bear in Hundredweights, as stated by Messrs. Newall and Co.

Girth in Inches.	lbs. Weight per Fathom.	Breaking Strain—Tons.	Working Load—Cwts.
1	1	2	6
1½	1½	3	9
1¾	2½	5	15
2	3½	7	21
2¼	4½	9	27
2½	5½	11	33
2¾	6½	13	39
3	7½	15	45
3¼	8½	17	51
3½	10	20	60
3¾	12	24	72
4	14	28	84
4¼	15	30	90
4⅜	16	32	96
4⅝	18	36	108

Note.—That the wire ropes, as manufactured, increase in their circumference by an eighth of an inch in each succeeding size of rope, but it has been thought sufficient in this table to advance by ¼ of an inch, and to omit the intermediate sizes.

The following scale shows the size, weight, and strength of flat wire ropes, and of equivalent flat hempen ropes, as stated by the same makers :—

Hemp.		Wire.			
Size in Inches.	lbs. Weight per Fathom.	Size in Inches.	lbs. Weight per Fathom.	Breaking Strain—Tons.	Working Load—Cwts.
4×1	$16\frac{1}{2}$	$2\frac{1}{8} \times \frac{1}{2}$	9	16	36
$4\frac{1}{2} \times 1\frac{1}{8}$	20	$2\frac{1}{2} \times \frac{1}{2}$	10	18	40
$5 \times 1\frac{1}{4}$	24	$2\frac{3}{4} \times \frac{5}{8}$	$12\frac{1}{2}$	$22\frac{1}{2}$	50
$5\frac{1}{2} \times 1\frac{3}{8}$	26	$3 \times \frac{5}{8}$	15	27	60
$6 \times 1\frac{1}{2}$	28	$3\frac{1}{2} \times \frac{5}{8}$	18	32	72
$7 \times 1\frac{7}{8}$	36	$4 \times \frac{5}{8}$	20	36	80
$8\frac{1}{4} \times 2\frac{1}{8}$	40	$4\frac{1}{2} \times \frac{5}{8}$	$22\frac{1}{2}$	40	90
$8\frac{1}{2} \times 2\frac{1}{4}$	45	$5 \times \frac{5}{8}$	25	45	100

It has been recommended that, instead of using a single chain or rope of large size, several parts of a smaller one should, by means of pulleys, be made to sustain the weight.

It is an advantage resulting from such an arrangement, that the crane or sheers may be used with greater facility for lifting smaller and more current weights than their maximum load, which it is not often needful to hoist.

Thus, by making the rope or chain form part of the machinery, it may not only be less ponderous itself, but toothed wheel-work of a lighter description may also be employed.

Referring to the engraving of the 10-ton crane (see Fig. 12), it will be observed that the occasional use of a single moveable pulley or "monkey block" in the bight or double of the chain, doubles the strength of the chain, and the power of the crane also; the power gained being equal to the number of moving parts of the chain or rope which suspend the weight, and are shortened as it is hoisted, or as the space through which the weight rises is to the length of rope hauled in.

The engraving of the 15-ton (see Fig. 13), shows a gain of 6 to 1 in power, by three pulleys in the lower block.

The upper block has four pulleys, but one them is for the purpose of leading the chain to the end of the jib, where it is made fast; so that this upper block may travel along the jib as upon a railway, without either hoisting or lowering the weight; and it will also be perceived that machinery is attached to the upper block, by which it may be made to traverse inwards or outwards, by a man who stands upon the ground, and hauls upon an endless chain acting on a sprocket-wheel above.

It is seldom the practice to gain more power by pulleys than 6 to 1, because it is inconvenient to place more than three pulleys, side by side, in the moveable block; but, when it becomes necessary to gain greater power with one pair of blocks, the pulleys are ranged in two heights, the upper tier of pulleys in the lower block being smaller than those in the lower tier, in order that the ropes or chains may work clear of each other.

In Mr. Armstrong's cranes, where great power is applied, in the first instance, by water-pressure, moving only through a short space, pulleys are used for the opposite purpose of increasing the height of the lift, power being lost in proportion to the distance through which the weight is raised, as compared with that through which the piston travels. (See Figs. 15, 16, &c.)

The author has observed that crane chains, in constant use, undergo a change in their internal structure; the iron, which was at first tough and fibrous, gradually becomes hard and crystalline; and he has then found them liable to break; he therefore recommends that every three years they should be taken down and heated to a bright red in the fire, and slowly cooled in ashes, or sawdust, to anneal the iron.

The Machinery of Cranes.

Wheels—their Proportion and Teeth—Cycloidal, Involute.—Professor Willis's Teeth.— Axles and Barrels, &c

In order that the power applied to cranes may be employed with the greatest effect, it is necessary that the wheelwork shall be properly designed and executed, otherwise power is expended to no purpose.

This power should be so proportioned to the work to be done, that, if manual labour be used, the men shall exert due disposable strength upon the work, at the speed of 220 feet per minute. If the power gained be too great, time is lost; if too small, the men's strength is overburdened, and eventually overcome; the work is too hard for them, and they must give it up. The radius of the winch, or handle, should not be less than 15 inches nor more than 18 inches, varying with the size of the men to be employed; 16 or 17 inches being the best average for ordinary labourers, and the height of the axle from the ground should not be less than 3 feet, nor more than 3 feet 3 inches. The best average is 3 feet 2 inches for the muscular exertion of a middle-sized man, and the pinion on the axle of the winch may have from 8 to 12 teeth. The pitch of the first motion-wheels may be $1\frac{1}{4}$ or $1\frac{1}{2}$ inch, and their width from 3 to $4\frac{1}{2}$ inches. The pitch and strength of the succeeding wheels must be proportioned to the stress which each has to bear, the stress increasing as power is gained by the wheelwork.

The power gained is thus calculated. In the 5-ton crane, the handles or winches have a radius of 17 inches, and the semi-diameter or radius of the barrel, measured to the centre of the chain rolled upon it, is $7\frac{7}{8}$, say 8 inches. The load, 5 tons, is equal to 11,200 lbs., and the wheelwork is as under, namely—

The first pinion, $4\frac{1}{2}$ inches diam., 11 teeth $1\frac{1}{4}$ pitch.
,, wheel, 3 feet ,, 89 ,, $1\frac{1}{4}$,,
Second pinion, 6 inches ,, 12 ,, $1\frac{1}{2}$,,
,, wheel, 4 feet ,, 96 ,, $1\frac{1}{2}$,,

$$\frac{\text{Barrel } 8 \times 11 \times 12 \times 11200 \text{ lbs.}}{\text{Winch } 17 \times 89 \times 96 \times 4 \text{ men}} = \frac{30800}{1513} = 20 \cdot 35 \text{ lbs.};$$

which is the statical resistance against each of the four men at the crane handles.

If this crane were constantly employed in lifting weights

of 5 tons, the machinery would not be sufficiently powerful, and the men would be overworked; but as it stands on a wharf to land and deliver general goods, which seldom weigh more than a ton, and rarely exceed 25 cwt., it is generally worked by two men, and the statical resistance to each man is from 10 to $12\frac{1}{2}$ lbs. When a load of 5 tons comes, the additional force of two canal boatmen enables the crane men readily to overcome a resistance of 20 lbs. for a single lift.

The packages, however, being chiefly about 4 cwt., the winch is transferred to the axle of the second pinion, and the first is disengaged; the combination is then

$$\frac{8 \times 12 \times 448}{17 \times 96 \times 2} = \frac{224}{17},$$

or 13 lbs. for each man.

The diameter of the wheels should be large, as compared with the axles and barrel, in order to avoid loss of power by friction, and the necks or journals should have sufficient length, say $1\frac{1}{2}$ times to twice their diameter, to preserve them from wearing or cutting, which is a mischievous expenditure of power.

The diameter and length of the barrel should be so proportioned to the lift of the crane that the length of the channel, cut like a square-threaded screw upon its surface, may be equal to the length of chain to be wound upon it, so that the latter part of the chain hoisted in may not ride or roll upon the first part; and the diameter of the barrel should, in any case, be such as to prevent any bending action or tendency to distortion of the links, as they apply their flat sides to the cylindrical surface of it, while the edge of every alternate link is received in the spiral groove.

If the edges or circumference of the pulleys be turned with a narrow groove in the middle, to receive the edge of the link while the flat of the alternate links rolls upon the pulleys, which should be turned, bored, and run upon a steeled or hardened pin, the chain will be prevented from twisting, and will work almost as smoothly as a leather belt.

The diameter of the toothed wheel upon the barrel-axle must be determined by the circumstances already stated, care being taken to make it as great as they may conveniently admit; the dimensions of the wheels of the 5-ton, 10-ton, and 15-ton cranes are good practical examples of proportion in similar cases.

In proportioning the diameters of axles, regard must be had to their length, to the diameter or radius of the wheel or lever to which the force is applied, and to the weight or stress at the end of the lever.

It is important that no practically sensible twist or torsion should take place in the axles, and it is found that the full force of four men (say 120 lbs.) may be exerted at winches having 18 inches radius at the ends of an axle 2 feet long and $1\frac{1}{2}$ inch diameter, without practically causing it to spring by twisting.

As a bar of iron subjected to twist is, in that respect, like a rope, the number of twists, and consequently the angle of torsion, will be directly proportionate to the length of the axle, to the radius of the lever that twists it, and to the weight or stress at the end of the lever.

And, as the springing of the axle by such twisting is radial deflection, the angle of torsion formed by the lever is, inversely, as the fourth power of the axle's diameter. The ultimate strength of axles is as the cubes of their diameters.

All pivots or journals subject to lateral stress should have their diameters in proportion to the cube roots of the weights upon them, and, for large pivots made of cast-iron, the cube root of the stress in hundredweights may be taken as the minimum diameter of the neck.

Thus, if one pivot or journal of cast-iron sustain a stress of 5 tons, or 100 cwt., the cube root is 4·64, and the diameter in inches must be $4\frac{3}{4}$ or 5 inches.

But, if the neck or pivot be of wrought-iron, the propor

tionate strength, as compared with cast-iron, is 14 to 9, or 100 to 64, the cube root of which is 4, the least diameter in inches which it is prudent to make a wrought-iron pivot, to carry 5 tons; as, for instance, at the end of a crane barrel: say, therefore, $4\frac{1}{2}$ in. These bearings sustain no twist.

Very few experiments have been made on the resistance of metals to torsion, especially on their elastic resistance. The best are those of Mr. George Rennie, who found that cast-iron bars, 1 in. square, were broken by a mean weight of 211 lbs. at a radius of 3 ft., which twisted the bar asunder close to the bearing, so that the broken bars may be considered as having no length. These experiments, in many respects interesting and valuable, are not applicable to the present subject; but, from the data here given, it is hoped that no difficulty will be found in assigning to each axle and pivot its right dimensions.

By the pitch of a toothed wheel is meant the distance from centre to centre, between two teeth measured upon the pitch line, which is the circle drawn through that point where two wheels, working together, come into contact with each other.

The widths of crane-wheels are from twice to three times as much as their pitch of teeth, that is to say, a wheel of $1\frac{1}{2}$-inch pitch is from 3 to $4\frac{1}{2}$ inches wide upon the face. The pitch is determined by the stress to be borne by the teeth; and it is important that every single tooth shall be fully capable of supporting the entire strain to which the wheel may be subject.

The strength of the tooth is calculated as that of a beam or lever fixed at one end and loaded at the pitch line, that is, about three-fifths or two-thirds of its length from the root of the tooth, where it may be said to grow out of the periphery of the wheel; for, although three teeth may be engaged at the same time, it is impossible that they can all be in perfect contact with their fellows, even with the best workmanship.

This is a point which should never be neglected or left to subordinates, who are too apt to take such models as the ironfounder may happen to have by him. The failure of a tooth may strip off others, and the wheel be broken, so that serious and fatal accidents may arise from such an oversight.

The next consideration is the form of teeth best adapted for crane-wheels; but it is to be observed that the thickness of the rim ought to be at least equal to that of the tooth, and be strengthened by a rib equal in section to a tooth; that the arms, at their point, should at least be equal in section to the rim; that they should be placed in the middle of the wheel, and be feathered on both sides.

Of the various methods which have been used to determine the forms of teeth for crane-wheels, the author has generally employed the epicycloidal curve produced by rolling a circle equal in diameter to the radius of the pinion upon another circle equal in diameter to the radius of the wheel, the diameters being taken at the pitch lines, which are the circles described by the wheel and pinion at their point of contact.

The curves so struck, commencing at the pitch lines, form the points of the teeth. They are struck in opposite directions, the space between their starting points being the thickness of the tooth; and from these two points radial lines were drawn to the centres of the wheel and pinion, which formed the sides of the teeth included between them, within the pitch line. This form, it will be observed, made the tooth smallest at the root by the convergence of the radial lines, and consequently tended to weaken it; this was remedied in the pinion by casting a plate upon the teeth, which, forming part of them, served not only to bind, as it were, all the teeth together, but to strengthen the body of the pinion, perforated and weakened by the axle passing through it.

"The roots of the teeth" upon the wheel were strengthened

by small angle pieces, for which space was found without the curved line described by the tooth of the pinion. Such teeth worked freely and equably together. But it will be observed that the side of each tooth of the wheel consisted partly of a radial line, partly of an epicycloidal curve, and partly of such a concave angle piece as might be found to clear the pinion; and it will also be observed that the wheel and pinion were adapted to each other; consequently another pinion, differing much in diameter from the first, would not act well, or, as a workman would say, pleasantly, with the same wheel.

Professor Willis, of Cambridge, has recommended a mode of forming the teeth of wheels by which this inconvenience is obviated. It was well known that the teeth of wheels, struck by the involutes of circles corresponding with their respective wheels, would work correctly with any other wheel of the same pitch made with involute teeth; but the obliquity of these teeth is often very inconvenient, although the writer has used them advantageously, when, from peculiar circumstances, as in rolling mills for making heavy bar iron, rails, &c., the wheels have, at times, more or less hold of each other, and the teeth work deep or shallow in gear. These teeth are struck by unwinding a string or ribbon from a roller, equal in diameter to the wheel, and describing the tooth by a tracing point at the end of the string.

The obliquity of teeth of the involute shape renders them useless for cranework, owing to the thrust caused by it, which tends to force the wheels asunder, and throws undue stress upon the axles and bearings.

The writer can state, from practice, the superiority of the form of tooth recommended by Professor Willis, which is thus produced.

If for a set of wheels of the same pitch a constant describing circle be taken to trace those parts of the teeth which project beyond each pitch line by rolling on the exterior circumference, and those parts which be within it by rolling on the interior circumference, then any two wheels of the set will work correctly together.

The describing or "*Pitch Circle*" should be equal in diameter to the *radius* of the smallest pinion, which, in this case, should not have less than twelve teeth. When rolled upon the interior circumference of a circle equal in *diameter* to the pinion, a point upon the periphery of the pitch circle will describe radial lines through the centre of the larger circle representing the pinion, which is twice the diameter. So that the form of the pinion teeth within the pitch line may be at once drawn in straight lines from the centre.

When rolled on the exterior circumference, epicycloidal curves, forming the teeth of the pinion beyond the pitch line, are described by the tracing point.

But, when these operations are performed by rolling the pitch circle upon another of much larger diameter representing the wheel, the interior and exterior epicycloids form a tooth of very different shape: it is no longer contained within radial lines, but spreads out at the root, giving great strength and firmness at the point where they are most needed. The exterior epicycloid forms the point of the tooth in a manner similar to that described in the first instance; but any wheel or pinion having teeth described by a common pitch circle will work together; even the teeth of a rack, which, being placed upon a straight line, may be regarded as the segment of a wheel of infinite radius, can be formed in the same manner, and will work equally well with the wheels.

Professor Willis has introduced another form of tooth, excellent for heavy machinery revolving always in the same direction, as the writer has experienced in practice, but not applicable to the wheels of a crane which work both ways round. To enter further into this subject would be to pass the limits of this work; but those who wish to do so may refer to Professor Willis's treatise on this subject, and to his paper in the Second Volume of the Transactions of the Institution of Civil Engineers. For the various other modes of striking the teeth of wheels they may consult Practical Essays on Millwork, edited by G. Rennie, Esq.

Professor Willis has also constructed an ingenious and useful instrument for striking the teeth of wheels with greater facility, which he has named "the Odontograph;" it is made both in card paper and metal, by Messrs. Holtzapfel, of Charing Cross.

Of the Foundations and Masonry for Fixing and Securing Cranes, and Reference to various Books and Authors for more complete and detailed Information.

Having thus briefly noticed the different parts which compose a crane, it is requisite to add a few words respecting the foundation for it, and the means of securing it to masonry, or buildings, or framework.

Referring to Fig. 39, which shows the foundation for the 10-ton crane, it will be seen that a mass of stonework forms the counterpoise to the suspended load; the strong cast-iron cross, into which the crane is stepped, lays hold of the masonry by means of the holding-down bolts and washer-plates, and, as it were, grasps the whole block. Reckoning the weight of the masonry, a light limestone, at the rate of 15 cubic feet to the ton, the diameter at 16 ft., and the depth 6 ft., the mass will weigh about 80 tons. The centre of gravity coincides with that of the crane-post, and the jib may be regarded as a lever, the fulcrum of which is placed under the end of the cross, 6 ft. from the centre of the mass. The sweep or radius of the crane being $19\frac{1}{2}$ ft., the arms of the lever are as $13\frac{1}{2}$ to 6 ft., the longest arm loaded with 10 tons, plus the projecting jib, pulleys, and chain, or about 2 tons more, gives the following statement, namely—

As 6 ft. : $13\frac{1}{2}$ ft. : : 12 tons : 24 tons,

the weight required to balance the crane.

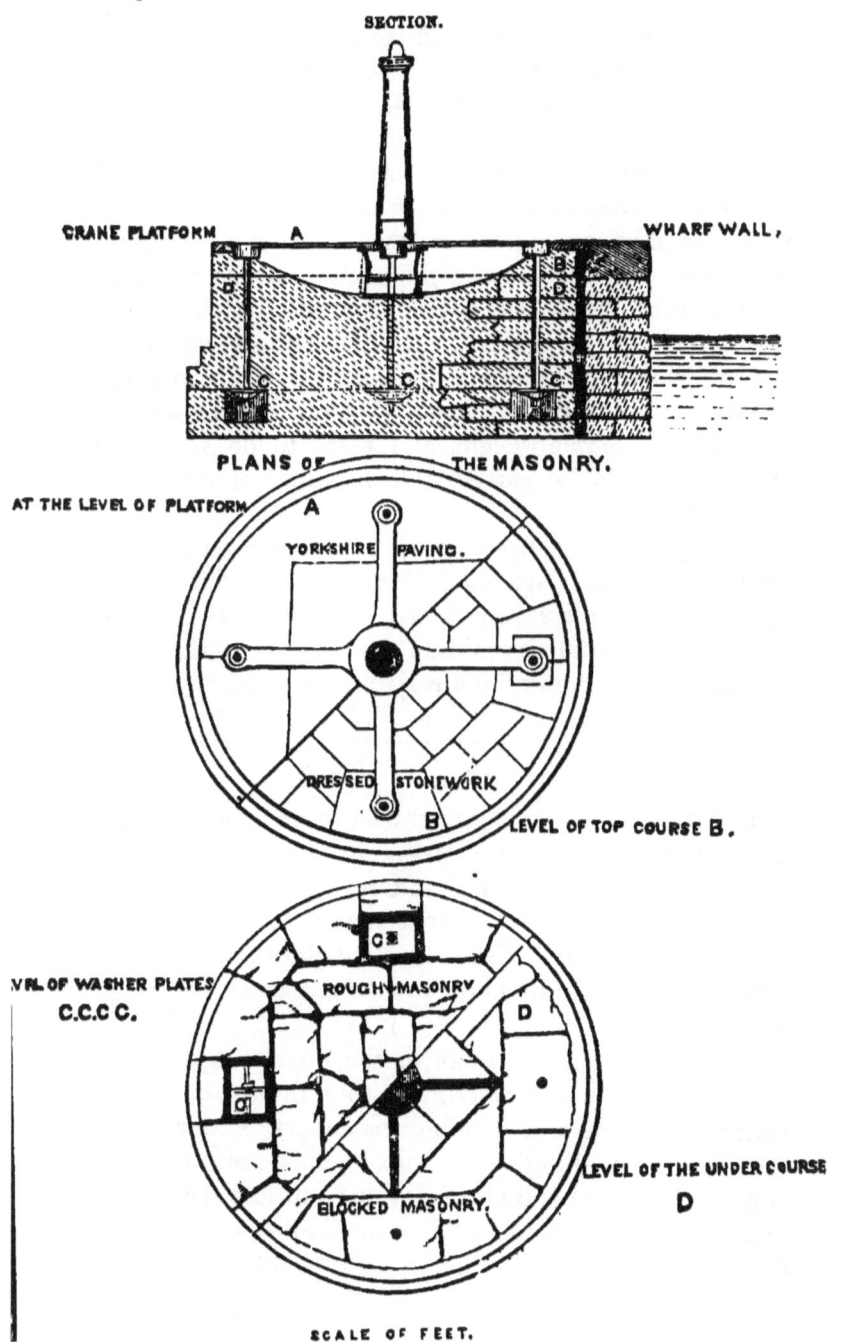

Fig. 39. THE MASONRY FOR THE 10-TON CRANE.

But the crane and its load must not only be balanced, but firmly held, without risk of disturbing its foundation; and, taking into account that the chain is sometimes carried beyond the perpendicular line, the weight of masonry, 80 tons, is not in excess. A marine engine boiler, wedged by its own weight into a narrow boat, or some similar oversight, sometimes doubles the load, and it is needless to say the crane is not calculated for such a strain; occurrences of this kind must be anticipated and provided for, or fatal accidents may ensue. In like manner, when cranes are attached to buildings, care must be taken that the resistance shall more than counterbalance the stress; and, if there be not ample weight and mass to do so, wrought-iron ties should be extended to lay hold of some further counterpoise.

Enough, perhaps, has been said on this point to warn the young practitioner against taking it for granted that buildings and roofs are secure, under such circumstances, without making careful and minute examination of the quantity, weight, and position of the resisting material.

For information on the manufacture of ropes and cables, he is referred to the "Professional Papers of the Royal Engineers," Volume 5th; and for their proportion and adaptation, to the calculations made in a very elaborate work, by John Edye, Esq., Assistant Surveyor of the Navy, entitled, "The Equipment and Displacement of Ships of War."

The "Professional Papers of the Royal Engineers" contain many important and interesting notices of the application of cranes and hoisting machinery, as, for instance, that of the travelling crane, with its tackle and framing for working the Diving Bell, in Volume 1st, and for the lifting and transport of heavy timber at Chatham Dockyard, by Sir Mark Isambard Brunel, in Volume 6th, besides many other cases, given in all their details throughout others of these volumes, which manifest the skill and judgment of the officers of that distinguished corps in their adaptation and use of mechanical means in the execution of works of utility as well as in military operations.

Buchanan's "Essays on Millwork," edited by George Rennie, Esq., and the papers of Professor Willis, already mentioned, are worthy of more especial notice; since it is rarely the case that persons so much occupied as these gentlemen are, either can or will devote their intervals of leisure to works on mechanical subjects.

The author has to express his thanks to several gentlemen connected with Government works for facts kindly communicated, and to the manufacturers of ropes and chain cables, who, without exception, liberally answered such inquiries as were made, especially to Joseph Crawhall, Esq., proprietor of St. Ann's Ropeworks, Newcastle-on-Tyne, whose extensive establishment contains a variety of ingenious mechanism, which it would be difficult to surpass, in the uniformity and excellence of the cordage it produces.

He is also greatly obliged to Messrs. Pow and Fawcus, chain cable-makers and anchorsmiths of North Shields, successors to Mr. Robert Flinn. The iron prepared by that gentleman for making harpoons for the Northern Whale Fishery has probably never been excelled in tenacity and strength; the shank of the barbed weapon was often bent into all forms by the exertions of the wounded whale, whose capture depended upon that slender rod of iron.

Messrs. John Abbott and Co., of the Park Iron Works, Gateshead-on-Tyne, kindly offered to make any experiments on chains which might be thought necessary; and Mr. William George Armstrong, of the Elswick Works, Newcastle, readily supplied drawings and details of his powerful water cranes.

This work has been written, in such intervals of professional engagement as presented themselves, from a wish to diffuse information respecting a class of machinery which the author has had constantly to employ and to construct for many years past; and also to second the views of the publisher, by contributing to a series of books for beginners, to be published at a cheap rate: for the first of these reasons the work will be found unequally written, and, for the last, it has been written almost in the same conversational manner in which he has been used to address his pupils and assistants.

REFERENCE TO THE ILLUSTRATIONS.

FIVE-TON CRANE.—This is shown by a Side-Elevation (Fig. 11), and a Back View (Fig. 11 a), and the mechanism may be described as follows:—
A, the principal Wheel, fixed on the barrel-axle, is in diameter 4 ft., and has 96 teeth, 1¼ in. pitch.
B, the Pinion, 6 in. in diam., 12 teeth, 1¼ in. pitch.
C, the Wheel on Second Motion, 3 ft.; 89 teeth, 1¼ in. pitch.
D, the Pinion, or Winch-axle, 4½ in.; 11 teeth, 1¼ in. pitch.
E, the Friction, or Brake Wheel.
F, the Barrel.
G, Guide-rollers for the chain.
H, the Collar, fitted with anti-friction rollers.
TEN-TON CRANE, also similarly shown (Fig. 12), has the wheel-work proportioned as follows:—
The principal Toothed Wheel, mounted on the barrel-axle, is in diameter 4 ft. 9 in., with 92 teeth, 2 in. pitch, and 4 in. broad.
Pinion, working in do., 7 in.; 11 teeth, 2 in. pitch, 4½ in. broad.
2nd Motion Wheel, do. 3 ft.; 72 teeth, 1½ in. pitch, 3 in. broad.
1st Motion Pinion, do. 6¼ in.; 12 teeth, 1½ in. pitch, 3¼ in. broad.
Machinery for turning the Cranes:—
Wheel on Crane-post, 2 ft. 8 in.; 52 teeth, 2 in. pitch, 3¾ in. broad.
Pinion working in do. 7 in.; 11 teeth, 2 in. pitch, 3¾ in. broad.
Bevel-wheel on same axle, 1 ft. 4½ in.; 32 teeth, 1½ in. pitch, 3 in. broad.
Do. Pinion working in do., 5¼ in.; 12 teeth, 1½ in. pitch, 3 in. broad.
FIFTEEN-TON CRANE (Fig. 13), with horizontal jib and traversing blocks, has the wheelwork the same as the ten-ton crane, but additional power is gained by the blocks, part of which is lost by the increased diameter of the barrel.
The machinery of the Traverse-Gear for moving the blocks is—
E, a Toothed Wheel, diameter 2 ft. 5 in.; 72 teeth, 1¼ in. pitch.
F, Pinion in do., diameter 4¼ in.; 10 teeth, 1¼ in. pitch.
Sprocket or chain wheel, 2 ft. 6 in., worked by an endless chain.
ENLARGED SCALE OF DETAILS of the fifteen-ton Crane shows a Side View of the Guide Roller; of the Traversing Rack, upon a portion of the

ON THE CONSTRUCTION OF CRANES. 119

Crane-Jib; a section of the Crane-Jib, Guide Roller, Traverse Pinion and Rack, with other parts of the Machinery of the Traverse-Gear for moving the Blocks along the Jib.

THE TRAVELLING CRANE (Figs. 14 a, 14 b, 14 c, 14 d) is shown by an End Elevation, Side Elevation, and Plan; and also by a view in Perspective, representing the Crane in work, as used for Stacking and Moving Heavy Materials.

THE WATER CRANES shown in Fig. 15, &c., are Two Self-Acting Machines, designed by Mr. Armstrong, and are copied from his Drawings. They represent a Wharf Crane, similar to that on the Quay at Newcastle-on-Tyne, and a Warehouse Crane, resembling those erected at the Docks in Liverpool, worked by the pressure of a column of water descending from a high Reservoir. The details of these are fully given in the text.

THE MASTING SHEERS at Her Majesty's Dock-yard, Woolwich, designed by the late Oliver Lang. Esq., the Master-Shipwright, are represented on a small scale (Fig. 32); but the following references give the dimensions of the Spars which are used to form these powerful Sheers—

A is the Mast, 44 in. in diameter, and 134 ft. long; it stands 114 ft. high above the Wharf, and is stepped 20 ft. into the masonry.

B, B, the Sheers, 26 in. in diameter, and 132 ft. long.

C, C, the Sprits, 24 in. in diameter, and 136 ft. long.

D, D, the Caps of the Sheers, 102 ft. asunder.

E, E, the Feet of the Sheers, 26 ft. apart.

F, F, the Wharf Wall 45 ft. thick at the top.

G, G, Wrought-Iron Braces.

H, the Flag-staff, 44 ft. long.

DIAGRAMS SHOWING THE STRESS on the different parts of a Crane. The various forms which may be given to Cranes, according to the purposes for which they are intended, or the situations in which they may be fixed, are shown in Figs. 34, 35, 36, 37, 38, and a simple mode of Measuring the Stress, by a Scale of equal parts, upon the Parallelograms into which the forces are resolved, is explained at length in the text.

FOUNDATIONS AND MASONRY for fixing and securing Cranes. Fig. 39 shows the Masonry used for the Ten-Ton Cranes as erected at Sleaford for the Canal Company, both in *Section*, showing the Crane-foot with its Cross, Holding-down Bolts, and Washer-plates; and a *Plan*, showing different Courses of the Stonework. A scale is attached, from which the parts may be measured.

INDEX.

Armstrong's water crane, 58.
Axle and block, 17.

Bar iron, strength of, 78.
Barlow's experiments on the strength of timber, 82; experiments on the transverse strength of wrought iron, 80.
Barrel axle of cranes, diameter of, 109; diameter of toothed wheel upon the, 109.
Barrels, crane, diameter and length of, 108.
Barrel crane for chains described, 97.
Beams, calculation of the strength of, 67; distribution of the load, 68; relative strength of round and square, 69; rules of the strength of, in different positions, 68; relative strength of, when laid edgeway, and flat, 67.
Beams, cast iron, form of, 70; proper load for, 72; deflection of, 81; sway, described, 6.
Beams, wooden, deflection of, 83.
Bellows, hydrostatic, 64.
Bits, windlass, 4.
Block, origin of the, 18; improvements on the, 19.
Block and axle described, 17.
Blocks, power gained by the use of, 19.
Block, monkey, increase of power by using, 105.
Brakes described, 33, 36.
Bramah's press, 64, 65; presses used at the works of the Conway Bridge described, 65.
Burgess and Walker's travelling crane, 47.

Cables, strength of, 98; table showing the weight and strength of chain and hempen, 103.
Capstan at the Woolwich dockyard described, 11; early form of, 8.
Capstan, improved, 9; improvements on the, described, 9.
Capstan, mine, described, 11; danger of, 11.
Capstan, "saucer" of a, 9; "step" of a, 9; use of, by the Russians, 14.
Capstans, working of, 11.
Capstan machine, prejudice of seamen against, 8, 15.
Cast iron, its resistance to compression, 74; modification in the structure of cranes by the use of, 33; tensile strength of, 74; tables showing the tensile, crushing, and transverse strength of, 77 and 78.
Cast iron beams, deflection of, 81; strength of, 70.
Cast iron pillars, strength of, 74; comparative strength of, with other materials, 76; Hodgkinson's calculations, 75; influence of the shape of the ends on the strength of, 75; strongest form of, 75.
Chain cable described, 97.
Chain and hempen cables, table showing the weight and strength of, 103.
Chain cables, proof strain for, in the Royal Navy, 101.
Chains, strength of, 98; comparison of the strength of, with ropes, 99.
Chains, crane, 97; testing ditto, 99.
Chinese, windlass used by the, 4.
"Cog and round" described, 17.
Colliery, Cinder Hill, application of steam power at the, described, 13.

G

Compression, resistance of cast iron to, 74.
Conway Bridge, Bramah's presses used in the erection of, described, 65.
Cordage, weight required to break, made by the old and new methods, 101.
Cornish man machine described, 52.
Cornish mines, use of the jackroll in the, 12.
Cotton mills, use of hoists in, 51.
Crab described, 8.
Crane, general form of, 92; stress upon each part of a, 92; strongest arrangement, 93; various modifications of the form considered, 93.
Crane beam, timber, strength of, 86.
Crane hooks, form and strength of, 81.
Crane posts, construction of, 36.
Crane wheels, size of, 106; diameter of, 108; teeth of, 110; form of teeth, 111; roots of teeth, 111; width of, 108.
Cranes, alteration in the form of, by the use of cast iron, 33; barrels of, for chains, described, 97; cast iron, proper load for, 72; change in the quality of iron of, by constant use, 106; chains for, 97; effect of brakes on, 72.
Cranes, fifteen ton, 38; five ton, 30.
Cranes, fixed, disadvantages of, 43.
Cranes, foundations for, 114; weight of foundation, 114.
Cranes, foundry, 41; goods, 28.
Cranes, origin of, 1; primitive, 3; progressive construction of, 26; shipwrights, 29; steam, 53; strut of, rules for determining the strength of, 88.
Cranes, ten ton, 34; travelling, 43; improved, 45; Messrs. Walker and Burgess's, 49.
Crane, vacuum, 53; walking, 27; water, 58; well, 36; wharf, described, 41.
Cross, power gained by the, 16.
Cylinders, relative strength of hollow and solid, 69.

'Dead-eye," origin of the term, 19.
Derbyshire lead mines, use of the jackroll in the, 12.
Derrick, Henderson's, 21; rigging a, described, 20.

Elastic strength of timber, 83.

Falconer, windlass described by, 4.
Field's experiments on the power of men, 23.
Fir, strength of a beam of Riga, 86.
Fixed cranes, disadvantage of, 43.
Flat ropes, 102.
Foundations for cranes, 114.
Foundry cranes, described, 41.

Gin, described, 12; limit of the use of, 13; use of, superseded by the steam-engine, 13.
Goods crane, 28.

Hague's vacuum crane, 53.
Hand wheel, 16.
Hempen cables, table showing the weight and strength of chain and, 103.
Henderson's derrick, 21.
Hoist, 51.
Hoisting machines, self-acting, 50; hoist, 51; lift, 51; man-machine, 52; Murray's lifting apparatus, 63; sack-tackle, 51; steam crane, 53; vacuum crane, 53; water crane, 58.
Hoisting machinery, application of, described, 20.
Hook's crane, form and strength of, 81.
Horse and gin described, 12.
Horse-power, origin of the term, 13.
Huddart's experiments on the strength of hand and machine-laid rope, 100.
Hydraulic press, 64, 65.
Hydrostatic bellows, 64.
Hydrostatic paradox described, 64.

Iron, change in the quality of the, of cranes, by constant use, 106.
Iron, bar, strength of, 79.
Iron moulds described, 41.
Iron, tables showing the tensile, crushing, and transverse strength of, cast, 77, 78.
Iron, wrought, strength of, 80.
Iron beams, cast, shape of, 70, Hodgkinson's calculations, 70; proper load for, 72; strength of, 70.
Iron pillars, cast, strength of, 74; comparative strength of, with other materials, 76; strongest form of, 75.
Iron pillars, wrought, strength of, 78.

INDEX.

Jack-roll, use of in the mines of Cornwall and Derbyshire, 12.

Kingston's experiments on the strength of bar iron, 78.

Lang's masting sheers, described, 90.
"Legs," the "three," described, 22.
Lift, 51.
Lifting apparatus, 63.

Machines, hoisting, self-acting, 50; hoist, 51; lift, 51; man-machine, 52; Murray's lifting apparatus, 63; sack tackle, 51; steam crane, 53; vacuum crane, 53; water crane, 58.
Machinery, hoisting application of, 20.
Man-machine, Cornish, 52.
Man, primitive state of, 1.
Masting sheers at Woolwich described, 90.
Material, deflection of, 81.
Men, power of, 23; table of the, 24; observations on the table, 25; Field's experiments, 23; Walker's experiments, 22, 25.
Mine, capstan, described, 11.
Mines, method of ascending from, 52.
"Monkey block," increase of power by using a, 105.
Moulds, iron, described, 41.
Murray's lifting apparatus, 63.

Odontograph, Willis's, described, 114.

Paradox, hydrostatic, 64.
"Parbuckle," described, 18.
"Paul and half-paul," 5.
Paul wheel, 6.
Pauls, description and use of, 4.
Pillar, timber, of cranes, rules to determine the strength of, 88.
Pillars, strength of cast iron, 74; influence of the shape of the ends on the strength, 75; shape of, 75; comparative strength of with other material, 76.
Pillars, wrought iron, strength of, 78.
Power of men, 23; Field's experiments, 23; Walker's experiments, 22, 25; table of the, 24; observations on the table, 25.
Prejudice of seamen against machine capstans, 8, 15.
Press, Bramah's, 64, 65.

Presses used in the erection of the Conway Bridge described, 65.
Primitive cranes, described 3.
Pulleys, advantage of groves in those used for chains, 108.

Reference, works of, 116.
Rennie's experiments on the resistance of metals to tortion, 110.
Rennie's travelling crane, 43.
Rope, effect of the quality of tar upon, 100.
Ropemaker's rule for estimating the strength and weight of ropes, 99.
Ropes, comparison of the strength of, with chains, 99.
Ropes, flat, described, 103.
Ropes, wire, 103.
Ropes, table showing the circumference, weight, breaking strength, and working load of, 104; table, showing the size, weight, and strength of flat, wire, and hempen, 105.
Round, cog and, described, 17.
Round beams, relative strength of, and square, 69.
Royal Navy, proof strain for chain cables, in the, 101.
Russians, extensive use of the capstan by the, described, 14.

Sack-tackle, 51.
"Saucer" of a capstan described, 9.
Seamen, prejudice of, against the use of machine capstans, 8, 15.
Self-acting hoisting machines, 43.
"Shammels," use of, by the Cornish miners, 12.
Sheers described, 21.
Sheers, masting, Lang's, 90.
Shipwright's crane, 29.
Smeaton's estimate of the power of men, 25.
Spokes, the origin of cogs, 16.
Steam crane, 53.
Steam power, application of, at the Cinder Hill Colliery, 13.
Step of a capstan described, 9.
Stress upon each part of a crane, explained, 92.
Struts, crane, rules to determine the strength of, 88.
Sway beams, described, 6.

Table of the circumference, weight, breaking strength, and working load of Newall's wire ropes, 104;

of the power of men, 24; of the strength of timber, 85; showing the size, weight, and strength of flat, wire, and hempen ropes, 105; showing the weight required to break cordage made by the old and new methods, 101; showing the weight and strength of chain and hempen cables, 103; of the strength of chains and ropes, 99.

Tables, showing the tensile, crushing, and transverse strength of cast iron, 77, 78.

Tar, effect of the quality of, upon ropes, 100.

Tensile strength of cast iron, 74.

"Three legs," described, 22.

Timber beams, deflection of, 83.

Timber, rules on the strength of, 86; elastic strength of, 83; strut of cranes, rules to determine the strength of, 88; table of the strength of, 84.

Tortion, Rennie's experiments on the resistance of metals to, 110.

Travelling cranes, 43; improved, 45; Rennie's, 43; of Messrs. Walker and Burgess, 49.

Trees, use of, as cranes, 2.

Triangle described, 22.

Vacuum crane, 53.

Walker and Burgess' travelling crane, 47.

Walker's experiments on the power of men, 22, 25.

Walking crane, 27.

Water, application of, for working cranes, &c., 58.

Water crane, described, 58.

Well cranes, described 58.

Wharf cranes, 36.

Wheel-hand, 16.

Wheel-paul, 6.

Wheels, crane, size of, 106; diameter of, 108; teeth of, 110; form of teeth, 111; root of teeth, 111; width of, 110.

Willis's improved wheel teeth, 112.

Willis's odontograph, 114.

Windlass, application of "paul and half paul," 5; improvements on the, 4; origin of the, 3; primitive, described, 3; use of iron paul-wheel described, 6; use of sway beams, 6; use of pauls, 4.

Windlass used by the Chinese, 4.

Windlass-bits, 4.

Wire ropes, 103; table of the circumference, *weight, breaking strength, and working load of, 104; table showing the size, weight, and strength of flat wire and hempen, 105.

Wooden beams, strength of, under various tests, 67.

Woolwich, masting sheers in the dockyard, described, 90.

Works of reference, 116.

Wrought iron, crushing, strength of, 78; experiments on the tensile strength of, 79; transverse strength of, 80.

THE END.

PRINTED BY JAMES S. VIRTUE, CITY ROAD, LONDON.

www.ingramcontent.com/pod-product-compliance
Lightning Source LLC
Chambersburg PA
CBHW020106170426
43199CB00009B/417